QUEEN OF THE OIL CLUB

QUEEN OF THE OIL CLUB

The Intrepid Wanda Jablonski

and the

Power of Information

Anna Rubino

BEACON PRESS, BOSTON

Beacon Press
25 Beacon Street
Boston, Massachusetts 02108-2892
www.beacon.org

Beacon Press books
are published under the auspices of
the Unitarian Universalist Association of Congregations.

11 10 09 08   8 7 6 5 4 3 2 1

This book is printed on acid-free paper that meets the
uncoated paper ANSI/NISO specifications for permanence
as revised in 1992.

Text design and composition by Susan E. Kelly
at Wilsted & Taylor Publishing Services

Library of Congress Cataloging-in-Publication Data

Rubino, Anna
    Queen of the oil club : the intrepid Wanda Jablonski and the
power of information / Anna Rubino.
        p. cm.
    Includes bibliographical references and index.
    ISBN 978-0-8070-7277-6
    1. Jablonski, Wanda, 1920–1992. 2. Journalists — United
States — Biography. 3. Petroleum industry and trade — Press
coverage — United States. I. Title.

    PN4874.J24R83 2008
    070.4'4933827282092 — dc22
    [B]   2007045468

Frontispiece: Wanda at her typewriter during her 1956–57 trip
to Saudi Arabia.

The final photograph herein, on page 271, appears courtesy
of Anna Rubino. All other photographs come from the Wanda
Jablonski Papers, a collection of Jablonski's personal and
professional records owned by Energy Intelligence. Reprinted
by permission of Energy Intelligence.

To Dick

and to our daughters,

Laurian, Caroline, and Elena

Contents

Foreword

There's a reason that, for so many years, people have spoken of oil*men*. During most of the history of the oil industry, though increasingly less true today, most of the people who made the decisions, whether in oil companies or oil-exporting countries, were men. And yet, over those many decades, there were two women who would have a decisive influence on the industry. Both were journalists, and neither hesitated to take on the most powerful.

One was the courageous Ida Tarbell, who at the beginning of the twentieth century published her history of John D. Rockefeller's Standard Oil Trust. That book has been described as the single most influential book on business ever published in America. It certainly helped energize Theodore Roosevelt to launch the great antitrust case that broke up Standard Oil in 1911.

But the other woman? She is much less well known. Her name? Wanda. This is not to be familiar; that's what people called her. After all, there was only one Wanda in the oil world, and that was Wanda Jablonski, who wielded enormous influence over the industry from the 1950s through the 1980s. Her life and her work have much to teach us about her era, oil and politics, and her own craft—and about independence and courage. It also provides a powerful perspective on today's critical questions about oil and energy, and about the complex relations between countries that import and countries that export oil and natural gas.

Wanda Jablonski was the real-time chronicler of the era when oil came to fuel the world economy, inflame nationalism, and ignite political turmoil. Moreover, as both an insider and an outsider, she had considerable effect on how it all turned out.

I first got to know Wanda in 1982 when we met at a seminar in a hacienda outside Mexico City. At that time, it was just becoming apparent that the price of oil, in the aftermath of the crises of the 1970s, was not going to spike to $100 a barrel as had been much predicted, but was more likely headed down. Professor Robert Mabro, who ran the Oxford Energy Seminar, gave a talk on what was changing, and Wanda realized it was news. She chucked her role as a seminar participant, and while the rest of us dutifully went back to our proceedings, she sat down in the courtyard with Mabro. There under the sun, smoking one cigarette after another, she worked with him, and as he later said, pushed him hard to turn his informal remarks into polished, pointed prose in time to meet her weekly deadline.

Some years later, I turned to her when I was researching my book *The Prize*. After first putting me off, she met me at the bar in the Carlyle Hotel, where she turned the tables and thoroughly interrogated me, and finally agreed. We began the first of what turned out to be several interviews, and then she took me back to her apartment, where she pulled out various papers and old documents. I remember that she was particularly touched by an article about her father, "Eugene Jablonski Returns to Botany," in *Garden Journal*. But it was her stories that really held me, about the oil ministers and the sheikhs, and about those who simultaneously were their collaborators and antagonists—the heads of companies and the tycoons. She had known them all.

I sought her out again to see if I could persuade her to go on television for the PBS/BBC documentary version of *The Prize*. Camera-shy and cautious about being interviewed publicly, she said no. Finally, Sue Lena Thompson, who worked with me on the project, sat up late into the night talking with her and at last won her over. But then, in January 1992, shortly before we were to film her, she suddenly died. It was a huge loss for history, and for the series—she would have stolen the show.

Now, in this compelling new biography, Anna Rubino tells Wanda's story, the whole story, which has never been told before. Rubino is uniquely qualified to do it: she worked in Wanda's newsroom in the 1980s and early 1990s and wrote her Ph.D. dissertation about Wanda at

Yale University under the guidance of the eminent historian John Morton Blum. Anyone interested in the Middle East, journalism, oil—or the changing role of women—will find *Queen of the Oil Club* compelling. But Wanda's story has much broader appeal. For here is a woman who, without riches or family connection, used her ingenuity, her perseverance, and her high standards as a journalist to challenge the status quo and to get people to see the world in new ways. Wanda's life story provides intimate insight into the movement that led to the oil crises of the 1970s and still affects today's geopolitical conflict in the Middle East, Latin America, and the former Soviet Union—the rise of oil nationalism. This book is an important contribution to the historical record because it tells, from Wanda's perspective, the story of how these oil nationalists came to challenge the seemingly all-powerful oil companies.

And what a story it is. The daughter of an itinerant Polish geologist and botanist, Wanda had, by the time she entered Cornell University, already been to school on four continents. She sometimes joined her father on geological expeditions, including one by camel in the Egyptian desert, after which she had to be deloused. From Cornell and then Columbia University, she went straight into journalism.

Imagine not being able to use your first name for the first decade of your professional career. Wanda Mary Jablonski had to hide behind her initials as "W. M. Jablonski" when she wrote for the *New York Journal of Commerce* because, in the post–World War II years, women writing about business were simply not taken seriously. By the late 1950s, however, Wanda had become so influential as a reporter for the journal *Petroleum Week* that she won the right to use her full byline, and by then was becoming known by her first name alone.

Yet she was still repeatedly shut out of industry gatherings held at men-only clubs, denied a visa for the Persian Gulf states because there were no "facilities for lady visitors" (until she pulled enough strings to get in), and invited to be the first woman to interview Saudi Arabia's King Saud—but only in his harem. In 1961, she took the virtually unprecedented step for a woman of starting her own publication—*Petroleum Intelligence Weekly*. Otherwise known as *PIW*, it was soon dubbed the "bible" of the oil industry. There was of course no Internet then, nor even a fax, and the newsletter's arrival each week by air mail on its yellow, parchment-like paper was, in varying mixtures, eagerly awaited and certainly dreaded. For it told in crisp, carefully researched stories what was going on in the world of oil—and that meant scoops. Wanda was a journalist to the bone, constantly cultivating sources and breaking news that people didn't necessarily want broken.

It wasn't easy. Almost everyone, at one time or another, seemed to suspect that she was a spy. People on all the different sides would get so mad at her that they wouldn't talk to her or, in the case of some of the companies, would cancel their subscriptions en masse—only to sheepishly resubscribe some months later.

Wanda was a truly audacious journalist who knew how to say that the emperor had no clothes. Other journalists were in awe of her, both because of her scoops and because her influence on the oil industry set new standards in business journalism for investigative reporting (even if the term itself was not yet in vogue). She was reporting on an industry known for its secrecy and for the entangling of politics and business in the service of national security, sovereignty, economic growth, and government revenues. She had to withstand many challenges—being grilled at the age of twenty-eight by the board of the world's biggest oil company, handling taunts from intimidating congressmen, avoiding jail time in Iran when suspected of being a spy. She was also particularly brave in dealing with challenges to her integrity and that of her staff.

Through her travels and reporting from the Middle East in the 1950s and 1960s, Wanda developed a perspective quite different from that of the Western oil industry, which led to stories and insights she would not have gained otherwise. She could also better see what was coming, for she grasped the consequences of the legacy of colonialist rule and the quest for sovereignty. She tried to communicate this in print and in person. "You never get beneath the surface," she chided the CEO of one of the Western oil giants. She advised him to "do yourself a favor" and get beyond "the red carpet treatment."

Anna Rubino has been a sleuth in pursuit of Wanda's story. She got exclusive access to Wanda's private papers, and she tracked down and interviewed more than a hundred people who knew Wanda, including former oil executives and oil ministers, Mobil Oil's long-retired chauffeur in London, a Catholic nun who remembered deep tensions between Wanda and her mother—and, of course, Wanda's colleagues, who knew what a demanding, difficult, and mercurial boss she could be. Rubino has searched through national archives on both sides of the Atlantic and even found Wanda's surviving relatives in Slovakia.

The result is *Queen of the Oil Club*, which shows a remarkable woman in action. This is a riveting life story that illuminates a whole era—and certainly helps make sense of our own times.

Daniel Yergin

Prologue: Wanda's Last Scoop

Wanda Jablonski had no intention of writing a story that evening in November 1986, when the telephone rang. She was trying to figure out who should succeed her as publisher of her prized *Petroleum Intelligence Weekly*. Long before Middle East oil became front-page news, Wanda—as she was universally known—had started this bible of the international oil world. For twenty-five years, her publication had been a must-read source of industry intelligence and a forum for dialogue between fierce competitors vying to control this strategic commodity. But she was sixty-six years old, and the time had come to pass it on.

Ensconced in a favorite red leather chair in her elegant penthouse on Manhattan's Upper East Side, Wanda was meeting with one of her editors, Tom Wallin, to discuss staff issues in private. At her request, he had brought along an old manual typewriter from the paper's midtown office to draft a letter about new organizational plans for *PIW*, as it was called. After years of making a living pounding out news reports on manual machines, Wanda refused to have an electric typewriter—or, heaven forbid, a computer—in her home. A legendary figure on the international oil scene since the 1950s, Wanda no longer wrote regularly about petroleum or the Middle East. Although she sometimes told riveting tales about the rise of the oil nationalists and their head-on clash with Big Oil, she shied away from describing her own unusual role in this drama. Besides, she still had a publication to worry about.

In her wood-paneled study, surrounded by treasures from decades of travel in the oil world—a hand-crafted chest from Kuwait, an ancient Bedouin saber and bronze dagger, an engraved copper platter and ewer—plus her collection of books, journalism awards, and photos of her posed with powerful men, Wanda presided like the chief executives and government ministers she had known personally over the course of four decades. A slender, petite woman with a short, stylish haircut, she had eased from brunette to honey blonde over the years, and her eyes and nose had benefited from plastic surgery. She favored slim, chic pants and prided herself on keeping up with the latest fashions. She had, some would say, aged gracefully.[1]

She was dictating her staff letter to Wallin when the telephone interrupted them. The phone had been ringing a lot recently, as the oil world was once again in turmoil. Prices, which had skyrocketed from $3 a barrel to $36 thanks to the oil crises of the 1970s, plummeted to $13 in 1986. The first massive oil shock, brought on by the 1973 Arab embargo, led to the largest transfer of wealth in history, the deepest worldwide recession in three decades, and a pervasive sense of energy insecurity in the West. Panic oil buying in the wake of Iran's 1979 revolution had produced the second steep spike in oil prices. But by 1986 it was the turn of the oil-exporting countries to be in shock, their fabulous prosperity imperiled, as higher crude oil output and reduced demand by consuming nations drove prices down much lower than experts expected. Not only were members of the Organization of Petroleum Exporting Countries (OPEC) in fiscal trouble, but so were non-OPEC oil exporters such as Britain and Mexico. Even worse for the Soviet Union, the price collapse threatened Mikhail Gorbachev's liberalization plans as revenue from oil exports shriveled.[2]

To Wanda's dismay, the first major victim of this worldwide oil glut turned out to be an old friend. In October 1986, Saudi Arabia's King Fahd fired his country's oil minister of nearly a quarter century, Sheikh Ahmed Zaki Yamani, OPEC's best-known leader. What's more, the Saudi royal family had put him under virtual house arrest. Although Yamani's troubles with the king had been brewing for some time, his abrupt dismissal—unceremoniously announced in a brief statement on late-night Saudi television, after the sports news—stunned the oil world. There was no explanation, no comment, not a word of appreciation for his long service to the oil-rich desert kingdom, the world's largest petroleum-exporting country. Unable to get a reaction from Yamani, the international press sought comments from his friends, including Wanda, but she honored their two decades of friendship by declining to go on the record. That was not her style.

The questions and speculations ranged widely. Was this a conflict over the kingdom's oil policy or an internal political power struggle? Did King Fahd blame his minister for plummeting oil prices? Was he upset with Yamani's recent public comments? Had Yamani's international stardom engendered so much jealousy in the royal family that the commoner had to go? No one knew for sure. "Many Westerners think they have insight into the Saudi system," one American academic expert on Saudi Arabia told the press, "but in reality Westerners know very little about it." Why Yamani was dismissed, the *New York Times* conceded a week after the fact, "is still veiled by Saudi secrecy."[3]

When Wanda answered her phone that November evening in New York, Wallin could tell at once that the call was important. She suddenly grew tense, listening closely, occasionally blowing wisps of smoke from her ever-present cigarette. Then her large hazel eyes narrowed, as they always did when she was acutely focused. Her tiny body, on high alert, seemed to coil.

She began to ask questions quickly, in staccato. Where was he calling from? Was the phone line secure? Who had sent him? The caller, a man she did not seem to know, had apparently just arrived in Europe from Saudi Arabia—London, Wallin guessed. It was about seven o'clock in Manhattan, the middle of the night for the caller. Abruptly, Wanda grabbed a few sheets of typing paper from Wallin and started to take notes in the large, looping script her staff knew so well. Soon running out of space, she signaled to Wallin: more paper. He quickly supplied it, but his puzzlement grew. It seemed as though Wanda, the renowned oil journalist and publisher, was taking dictation. Was she transcribing an official statement? Wanda read back to the mystery caller what she had written, and asked, "Is that correct?" She took more notes and again read them aloud, word for word, to her caller. When she finally hung up, her eyes gleamed. "This is great!" she exclaimed. But who was it? She would not say.[4]

The next day, Wanda burst into her publication's Times Square headquarters, the triangular ticker-tape building on Forty-Second Street, and presented her most senior editor, Ken Miller, with a draft article and her notes. "This is the biggest story," she exclaimed. She declined to tell him the source but ordered him nonetheless to use the information word for word. He was not even allowed to verify it with other contacts. Miller, a quiet, soft-spoken man in his early sixties, arched his eyebrows as he listened and then read her story. He was stunned—and skeptical, both about the news and about Wanda's directive. A Middle East oil correspondent for some three decades, Miller was suspicious by profession and by temperament of any new information from that region. And this order was so unlike Wanda. She always insisted on the highest standards of news gathering; everything had to be checked and double-checked. One-source stories were exceedingly rare. What's more, the partially drafted article she gave Miller to edit was a bombshell. It could jeopardize the entire publication. *PIW*'s reputation was at stake.

Miller knew, of course, that Wanda was a close friend of Yamani. As chief spokesman for his country's oil policies since 1962 and the preeminent leader in OPEC, Yamani became widely known in the West during the oil crises of the 1970s through frequent television appearances and front-page photos from OPEC conferences. With a distinctive goatee, liquid dark eyes, and a wardrobe that moved easily between natty pinstripes and traditional Arab robes, he was the public face of Arab oil. Wanda, however, had befriended Yamani long before his rise to fame, back when he knew little about the inner workings of the oil business and most of the Western press ignored him. Her access to the Saudi oil minister was exceptional.[5]

The article Miller would edit described Yamani's "final showdown" with King Fahd and the secret reasons for his dismissal and confinement to his home. The story deeply embarrassed the royal family because it revealed duplicity at the core of Saudi oil policy. Years later, in an interview about his nearly thirty-year friendship with Wanda, Yamani said that the article had worried Saudi leaders so much that they had felt obliged to release him—lifting embarrassing travel restrictions, returning his passport, and, most important, halting their surreptitious efforts to smear his name.

"Wanda stopped the royal family," Yamani said, snapping his fingers, "just like that."[6]

Wanda Jablonski was once the most powerful woman in the world of oil. She owned neither petroleum deposits nor drilling platforms, refineries nor supertankers. As an investigative reporter and publisher, what she had was information: exclusive, highly prized intelligence on an industry that thrived on secrecy. She—and she alone—had extraordinary personal influence with key players on both sides of the titanic struggle between oil-producing and -consuming countries that led, in the 1970s, to a massive, global economic upheaval. The major oil companies, those industrial giants of the twentieth century, once relied on the complexity and confidentiality that shielded them from public scrutiny. By lifting their veil of mystery, revealing many of their strategies and vulnerabilities through her reporting and her informal intelligence network, Wanda wielded exceptional clout.

Her article about the showdown between Fahd and Yamani in late 1986 turned out to be the final, signature piece of a career that spanned the four decades of the cold war and the emergence of the Middle East as a critical geopolitical region for the United States. When Wanda died in 1992, Pulitzer Prize–winning historian Daniel Yergin acclaimed her to be one of "two great women journalists who had far-reaching impact on the world oil industry." Nine decades earlier, in 1904, Ida Tarbell had written her famous muckraking history of the world's most powerful multinational corporation, the Standard Oil Co. Just as Tarbell spurred the trust-busting of that company during Theodore Roosevelt's presidency, so Wanda Jablonski contributed to the dismantling of the postwar international oil club in the 1970s. As Yergin put it, Jablonski "not only chronicled but also helped shape the postwar oil industry."[7]

The members of this elite club—engineers, geologists, and accountants who managed the seven largest petroleum companies—were fiercely competitive in their quest for profit and power, yet they had much in common. They were all white, all Western, all men. Through formal gatherings and informal encounters, they controlled access to their ranks and set the basic parameters for their businesses by establishing and maintaining a web of joint ventures around the globe. Until the 1970s, their overlapping agreements and mutual interests had maintained relatively stable oil prices and sufficient production for several decades, despite worldwide economic crises and wars. It was this mastery of oil—the most prized, most lucrative commodity in the industrialized world—that made these seven companies virtually self-sufficient. Most oil-producing nations had ceded their ownership rights in exchange for royalties. If oil companies could find the crude oil, it was theirs to develop and sell as they pleased.

And find it they did, especially in the Middle East. By the 1950s, the executives of the so-called Seven Sisters—a group of American, British, and Dutch oil companies that included Exxon, British Petroleum, and Royal Dutch/Shell—controlled nearly half the world's total trade, managed incomes larger than those of many nations, and imposed their decisions on bankers, bureaucrats, and politicians. And as the president of Conoco, which was not one of the seven, noted in 1967, it was an exclusive group. "When you're breaking in, you're sons-of-bitches: then you become a member of the Club," he said. "We hope to join as the Eighth Sister."[8]

Into that club, uninvited, stepped Wanda Jablonski, a Slovak-born American journalist who shook up these leaders of the Seven Sisters. Her scoops and commentaries on their industry—written for a business newspaper, magazines, and, later, her own industry newsletter—were unlike any they had ever seen. This book is the story of how a woman gained influence over that informal fraternity in the 1950s and 1960s—the years of its greatest control over the international oil trade—and how she contributed to the breakdown of its power by exposing its weaknesses and informing its opponents. Her story is also a primer on the recent history of oil, reminding us how the quest for stable sources of oil became critical to American security, and how nascent nationalism in the 1950s led oil states in the Middle East to use petroleum as a political tool. As the most influential woman on the oil scene, Wanda reversed the typically male gaze to fix her eyes on these men, and she mastered their world.

Oil has been central to the American way of life since the mid-twentieth century. After World War II, as inexpensive domestic production fueled burgeoning economic growth, oil displaced coal as the principal source of power in the United States. The country's appetite for oil tripled from 1948 to 1972, exceeding domestic production and making access to secure foreign supplies a national priority. More recently, oil has played a significant role in two U.S. wars in the Middle East. Despite the rhetoric of national sovereignty, weapons of mass destruction, and spreading democracy, what is really at stake is access to the region's vast oil reserves. With petroleum imports now supplying more than half of U.S. consumption—compared with less than a quarter in 1970—Americans today cannot live without Middle East oil.

Western mastery of the petroleum world was once a much simpler affair. In the years after World War II, executives of the Seven Sisters (also known as "Big Oil" and "the majors") took care of the details. Handling some 90 percent of world petroleum production and trade, this elite group of industrialists assured the world of a steady stream of oil for more than a quarter century, with minimal government involvement. But as Western countries became more dependent on oil from developing nations, particularly the Middle East, the majors' way of doing business came increasingly under fire in the oil-producing countries.

The challenge came from oil nationalists, who advocated government control over their own countries' natural resources. At stake were both money and power. In the first half of the twentieth century, a tenant-landlord relationship developed between the companies (the tenants, who risked capital and manpower to find and develop the oil) and the oil-rich developing countries (the landlords, who collected the rent). But the tenants held the upper hand, as their contracts (called "concessions") with the landlord governments gave them control over production rates and prices. Outright nationalization was problematic because the oil club knew how to retaliate. The majors successfully organized boycotts against Mexico, which nationalized its oil industry in 1938, and then Iran in 1951. In the 1950s and early 1960s, however, unexpectedly prolific discoveries in the Middle East and North Africa intensified frustrations in the landlord countries, which depended heavily on oil royalties in their national budgets. When the majors cut oil prices, and hence the oil countries' revenues, twice without consultation in 1959 and 1960, nationalists from the Middle East and Venezuela persuaded their governments to take action. In 1960, they joined forces to create OPEC—a club to counter the oilmen's club, a cartel to challenge what they considered a corporate cartel.

Excess oil supplies thwarted OPEC's initial efforts to exert greater control, but when demand outpaced supply in the early 1970s, OPEC countries checkmated the oil club by pitting one company against another. Thanks to this newfound power, the Arab members of OPEC then used oil as a political and economic weapon, which led to the oil crises of the 1970s and the massive economic shocks that ensued. This was the struggle that Wanda Jablonski entered, and it was there that she left her mark.

The Middle East and the oil industry might seem difficult subjects for a female reporter, given how male-dominated both of these worlds were and still are. Indeed, fifty years ago, the idea of a woman covering the oil industry seemed absurd. "Girl reporters just couldn't cover oil," one chief executive remembered thinking in 1945 when he met Wanda, a young reporter for the *New York Journal of Commerce*. Reporting from the Middle East in the 1950s, Wanda wrote to her colleagues that "people seem so amazed at meeting a 'lady oil expert' that I'm beginning to feel peculiar, like I had two heads or something."[9] Few women journalists at that time wrote for anything beyond the women's pages. Business or economic stories under a woman's byline held no credibility. Ironically, given that she came to be known in the oil world solely by her first name (and hence I use her first name in this book), Wanda felt obliged to hide her sex for more than a decade. She used her initials, "W. M." for "Wanda Mary," in her byline until she was bold enough to start using her first name regularly in 1957.

Wanda flouted other conventions, too. Without family money or a powerful mentor, she became the first woman to create a major business publication in the United States. When faced with the loss of her job in 1961, Wanda founded *Petroleum Intelligence Weekly* and charged the then-outrageous subscription price of $365 a year—just a dollar a day, she would say. To her it was a "terrible gamble," but the newsletter soon became the industry's prime source for exclusive intelligence and an important forum for debate. When petroleum rose to front-page status in the general press during the economic crises of the 1970s, *Petroleum Intelligence Weekly* became an indispensable source of news for a broader audience and made her a multimillionaire.

Wanda Jablonski's success as a reporter and then a power broker derived in part from her unusual position as both an insider and an outsider. Born amid the ravages of Central Europe in 1920, she grew up an itinerant, following her oil-geologist father around the world from one exploration hot spot to another. She learned about oil through his eyes, but she mastered the ins and outs of the oil market on her own. Working her way up from messenger ("copyboy") to cub reporter on the *New York Journal of Commerce*, Wanda cultivated an uncanny ability to coax information out of people. As an oil expert, she knew the world of derricks and drilling; she gained and kept the respect of the industry's leaders. But as a woman and as a reporter, she remained on the periphery of that world, maintaining and protecting her own distinct perspective.

Particularly in the Middle East, this balance between being an insider and an outsider made a difference. On reporting trips in the 1950s and 1960s, first for the *Journal of Commerce* and then as international editor of the McGraw-Hill magazine *Petroleum Week*, Wanda grew concerned about the inherent inequity in the West's control of world oil production. Although deeply anticommunist and a proponent of market capitalism, she came to question and finally challenge the rules of the game. She began to look at the world through the eyes of those whose countries had the petroleum but not the power, who benefited far less from their oil than the industrialized West did. These countries had little say over what was done with their natural resource, including how much was produced and where it was sold. Wanda came to believe that although the West needed to maintain influence in the Middle East because of its prolific oil reservoirs, some of Big Oil's methods were wrong. And so she became one of the first to identify and publicize the ideas and arguments of the lawyers and engineers who became the leading oil nationalists. She also helped them, both publicly and privately, by revealing fiercely held secrets of the international oil business, especially the intricate ways in which the majors adjusted production levels and prices. Though no muckraking reformer, Wanda Jablonski, like Ida Tarbell, showed that the press, by exposing and explaining business practices, could exert considerable influence.

Wanda did not create these oil nationalists—OPEC would have challenged this club of international oil companies without her—but she is rightly regarded as the midwife of OPEC's creation. She recognized the importance of its cofounders, Venezuela's Juan Pablo Perez Alfonso and Saudi Arabia's Abdullah Tariki, well before they met each other. Though the Western business press often demeaned their efforts in the 1950s, her substantive coverage gave them credibility, forcing the oil companies to pay more attention to their arguments. She privately talked with the nationalists about arcane trading arrangements, secret rules of the game, and why they should work together to gain greater leverage. Although Tariki and Perez Alfonso were planning to meet at the historic Arab Petroleum Congress of 1959 in Cairo, it was Wanda who made the introductions—over Coke or bourbon, depending on who was listening to her story—a typical, quietly influential moment that would change the nature of the oil trade forever.

Wanda's pursuit of oil news led her in the 1950s to places where few, if any, Western journalists had ventured, let alone women reporters. From the Persian Gulf's "Pirate Coast," where a sword-wielding sheikh asked about the quality of grazing in her hometown of New York, to an oil camp in a Venezuelan jungle, where the roustabouts worked under threat of attack by arrows from Motilone natives, she explained and humanized oil-producing countries and international exploration work to her desk-bound readers—Western oil executives, bankers, and government officials.

Wanda Jablonski broke through as an investigative reporter in three exceptionally macho postwar worlds: the international oil industry, the Middle East, and business journalism. Her boldness, glamour, and legendary exploits make for a lively tale, but her story transcends its time and place. Wanda showed that information is power, that women can excel at getting and analyzing insider information, and that the ability to see the world from another culture's perspective is critical. The history of international oil development informs decisions made today about American strategy on Iraq, Iran, Saudi Arabia, and the other OPEC nations. Wanda's life adds texture, context, and color to the economic and political forces that still pit Western consumers against Middle Eastern oil producers.

A number of Wanda's peers believe that no other business reporter in the second half of the twentieth century had as much influence over an industry as Wanda had over the international oil industry. As a reporter and then a publisher, Wanda helped spell out the agenda of the day for both the oilmen's club and OPEC, whether they liked her or not. Despite attempts to boycott her publication and to brand her a spy, she maintained her integrity and prospered. Through her scoops and commentaries and then her own weekly publication, she focused the industry's attention on key issues, especially the oil nations' desire for greater participation in the actual business of oil. Her status as an insider gave authority to her perspective as an outsider. As a Royal Dutch/Shell senior executive, Howard Macdonald, said, "Information is power, and Wanda had it in spades."[10]

This book is the first major account of Wanda Jablonski's life, based on exclusive access to her private papers and other private collections, newly declassified government documents, and more than a hundred interviews with oil-company senior executives, oil ministers, OPEC officials, journalists, and family members. Although Wanda did not keep a diary, her voice still emerges from her published articles over four decades, her small collection of letters and memoranda, and the recollections of those who knew her. Because her job became her life, her personal life—her marriage, her occasional affairs, and her friendships—suffered from her single-minded drive to be the best reporter on oil. Despite the lack of information about her inner life, it is her work as a journalist that serves as my prism for the world of oil in transition.

When I worked as a reporter at *Petroleum Intelligence Weekly* for four years in the 1980s and for an affiliated publication for two years in the 1990s, I knew Wanda and traveled with her once for interviews in Mexico. I admired her high standards, but by then she was a difficult, irascible boss. At her death in 1992, I helped prepare a special edition of *Petroleum Intelligence Weekly* on her legacy. I discovered that the seemingly implausible stories she had once told me and my colleagues were, for the most part, true. I was also struck by testimonials from former oil-company presidents, government officials, and oil ministers that underscored just how much of a trailblazer she had been in the oil world and in business journalism. By defying conventions in a notoriously male-dominated time and place, by questioning and probing and refining her own perspective, she helped change parameters that had seemed fixed. Using only her wit and her words, Wanda got people thinking in new ways. That alone was enough to get me to tell her story.

1.

Peripatetic Youth

The medieval city of Trnava, in the foothills of the Low Carpathian Mountains in western Slovakia, was known during its heyday as "Little Rome." In the sixteenth century, huge, fortified walls and a pious population made Trnava an ideal refuge for the Hungarian archbishop of Esztergom, who had been forced to flee his bishopric to escape the invading Turks. For the next two hundred years, the city flourished as the ecclesiastical center of the region and the home of a growing Catholic university. A cathedral, numerous churches, and prosperous homes with Gothic, Renaissance, and Baroque façades still attest to the city's period of greatest power, but by the eighteenth century, Trnava's prominence was eclipsed. The nearby city of Bratislava, more strategically located on the Danube River, took over as the economic, political, and religious center of Slovakia. "Little Rome" turned into a quiet provincial burg.

Wanda Mary Jablonski was born in Trnava on August 20, 1920. At that time, Poland's northern border was under attack from the Soviet Red Army, and it was no place to have a baby. So Maria Krcmery Jablonski, pregnant with Wanda, traveled alone over the mountains in the spring of 1920 to find refuge with her family in Trnava while Wanda's father, Eugen Jablonski, remained in Poland, where he was struggling to make a living as a geologist for a government survey team. Maria joined her older brother, Karol Krcmery Jr., a prominent surgeon and university hospital director, who lived with his wife, Mahulena, and their young children in the director's residence on the hospital grounds. This rather grand house, surrounded by a handsome rose garden, became Wanda's first home, but Maria decided only a short while after giving birth to rejoin Eugen in Poland. Her concerns for her husband outweighed the immediate needs of her infant, their only child. She left Wanda at Karol's house in the care of a nursemaid.[1]

Although calm compared with Poland, Slovakia in the summer of 1920 was a restless, anxious land. The anarchy and political upheavals that followed World War I left the region exhausted and groping for a sense of identity. Slovakia had survived centuries of Habsburg rule by maintaining, however tenuously, its own language and cultural heritage. Gradually the region's prospects revived, and with the collapse of Habsburg control, Slovakia and the neighboring Czech province established in 1918 the Czechoslovak Republic. By the time Maria Jablonski arrived in Trnava in 1920, many Slovaks hoped their culture, finally freed from Austro-Hungarian domination, would flourish once more.[2] Although Wanda spent only the first few years of her life in Trnava, her birthplace and family history nurtured in her an appreciation of nationalist concerns that would influence her work as a journalist.

NATIONALISM IN THE BLOOD

"In order to understand my background," Wanda Jablonski explained in a 1976 interview, "you must start with an understanding of my grandfather."[3] Karol Krcmery Sr., Wanda's maternal grandfather, was a fierce Slovak nationalist. A deeply religious man, born in 1859 under Slovakia's snow-capped High Tatra Mountains, Krcmery earned his doctorate in linguistics at the University of Vienna. To find a job, he migrated to the Serbian town of Kikinda and taught Latin and Greek in a gymnasium, a secondary school, where he eventually became the headmaster. He married a woman of Armenian descent, Anna Zakarias, from whom Wanda inherited her auburn hair. Anna bore seven children, and five survived to adulthood. In 1906, while caring for a schoolmate suffering from tuberculosis, Anna succumbed to the disease and died, as did two of her daughters. Devastated, Karol Krcmery sent Maria and her siblings to a nearby village for safety. Sometime later, he brought back his other children to live with him in Kikinda, but he sent Maria, then seventeen, to live with a family in Budapest. A promising student, she would finish her secondary studies there and attend the university.[4]

With the empire's defeat in 1918, Krcmery returned to his native Slovakia, an agricultural land dwarfed by the more economically robust, politically adept Czech region. Repression under semifeudal Austro-Hungarian rule had led to mass migrations of Slovaks to the United States. Widespread poverty and illiteracy hampered efforts to stand up to the Hungarian, German, and Czech elites that dominated the region, and religious differences compounded local resentment: the ruling class was mostly Protestant, whereas the vast majority of Slovaks was Catholic. In late 1918, Krcmery joined with a number of Slovak Catholic nationalists, including the flamboyant priest Andrej Hlinka, to organize a group dedicated to Slovak political, cultural, and linguistic autonomy within the Czechoslovak federation. He helped reestablish a formerly banned Slovak cultural institution and codify the Slovak language, which could then be taught in school again. Elected in 1925 to the National Assembly, Krcmery eventually became the leader of the Hlinka party in the Czechoslovak Senate.[5]

Senator Krcmery's fierce nationalism was imprinted on his granddaughter, Wanda, from her earliest days. For his daughter Maria, however, Krcmery's passion for politics was unsettling. Born in Kikinda in 1888 and educated in Budapest, Maria had never lived in Slovakia. To her family's consternation—in addition to her father, her brothers and cousins were also ardent Slovaks—Maria considered herself Hungarian and deeply resented the postwar breakup of Hungary. In an unpublished essay about her husband's career, written in the 1960s, she blamed Woodrow Wilson for her estrangement from her family: "You have no idea what Wilson's fourteen points did to Hungary. A nation that managed her minority problems for more than one thousand years, now with the introduction of the idea of self-determination, the country went to pieces. Families were torn apart, the whole country disintegrated. My own family is one of the saddest examples." Maria's Slovak relatives saw it differently. Dr. Vladimir Krcmery, one of Wanda's closest cousins, said that Maria was the only one in the family who felt estranged.[6]

Wanda remembered her grandfather being scornful of Woodrow Wilson, not for Hungary's dismantlement, but for conceding too much to the Czech nationalists who rejected Slovak demands for autonomy. In a 1976 interview, Wanda recalled that in moments of stress, her grandfather would utter what she thought was a Slovak expletive, "Vooderveels." Only later did she find out that he was saying, with great contempt, the name of the American president. Thinking back on her early years in Slovakia, Wanda remembered her grandfather as an austere, autocratic paterfamilias who played politics, manipulated Catholic priests, and conversed regularly in Latin.[7]

The Trnava house was a grand, lively mansion, where musicians would sometimes play in the evening, but Wanda was lonely there. She felt small and fragile, both physically and emotionally. In 1990, on her first return to Trnava since her teen years, "she told me she was not very happy in this house," said her cousin Vladimir. "She felt [her] parents put her there because they wanted to make large travels and it was not possible with such a [young] child." She felt oppressed by her older cousins. Family photos of this time include one of a serious-looking Wanda wedged into a wooden cart with another girl, being pushed and pulled by two larger boys.[8]

MARIA AND EUGEN

Wanda's mother, Maria Krcmery, must have also felt pushed and pulled. Losing her mother and two sisters at such an early age, she, too, must have felt abandoned as a child, particularly after her father sent her to study in Budapest. She met Eugen Jablonski, a fellow student at the Hungarian Science University, in 1910. Already a mathematics teacher with two diplomas, Maria was one of twelve high-school instructors selected to pursue an additional degree. She chose botany as her main subject, as had Eugen. "One day while working in the Assistant's room," she wrote, "I heard them discussing the students: Jablonski—what a mind! There never was one like his in a hundred years. 'Who is he?' I asked. 'Please point him out.'" He was a slender,

Wanda at about the age of four at her uncle's home in Trnava.

wiry young man of modest height with angular features and a wide smile.

Eugen was "completely absorbed" by botany, Maria quickly discovered. Unfortunately for her, he was drawn to the great German universities, initially Breslau and then Berlin, where he contributed in 1915 to a major academic study on plants. Shortly thereafter, at a meeting of the Hungarian Botanical Society, Maria heard some fellow botanists say that they "did not have enough adjectives to praise" his work, but they "expressed regret that he hadn't acknowledged himself to be a Hungarian." She knew that, in his heart, Eugen was a Pole. He called himself a *goral*, meaning "mountain folk" in Polish.[9]

Though not as obsessed with nationalism as Senator Krcmery was, Wanda's father had his own difficulties with national identity. Eugen, like Maria, was the product of a multiethnic blend. Born in 1890 to a Polish father and a Hungarian mother, he grew up in Eger, in eastern Hungary. His father, Florian Jablonski, the son of Polish farmers from the Polish-Slovak village of Jablonka in the Carpathian Mountains, had moved to Eger to teach botany and geology. As a child, Eugen accompanied his father on weekend botanical trips, so that by high school he knew the native plants and flowers by their Latin names. Eugen would later do the same with Wanda.[10]

After completing his work in Berlin, Eugen returned to Budapest to finish his doctorate in botany, with a minor in geology. Happier in the field than in the classroom, he turned down university positions to join a geological expedition in 1914 to a remote region of Hungary. When the scientists returned to "civilization," as Eugen later recalled, they discovered that war had broken out. They were arrested as Serbian spies because of their odd-looking equipment. When they were released, Eugen was drafted and packed off, after only two weeks' training, to the Russian front. Like so many nonethnic Hungarians who lived in Hungary, this intense young botanist could not escape conscription to an army to which he had no loyalty.

With the Austro-Hungarian defeat at Podolia in September 1915, Eugen was taken prisoner and kept for three years in the Tsaritsyn camp on the Volga. His sweetheart, Maria, managed to send him some packages, including books on geology that he studied "from cover to cover." Then, in 1918, amid the chaos in Russia, Eugen and a friend escaped: they jumped a prison wall, but they could not go far because the friend injured his leg. Thanks in part to Eugen's ability to speak some Russian, a Cossack family in a nearby village took them in. The next evening, while waiting for his friend to recuperate, Eugen crept back to the prison to alert others that the Cossack family would help. This episode was "very characteristic of my husband, as a *man*," Maria wrote. "He was risking his life, as he could have been shot or he could have been captured again. His coming back to Budapest is an odyssey alone."[11]

By the time Eugen rejoined Maria in Budapest in 1918, Austro-Hungary was about to lose the war. Eugen managed to get a job as curator of tropical plants at the Botanical Gardens, and Maria continued to teach. Shortly thereafter, the two were engaged. "After nine years of acquaintanceship, courting, working together, kin folks were urging us to get married," Maria recounted. By then, in the spring of 1919, Hungarian communists, led by another former prisoner of war, Bela Kun, had taken control in Budapest. When the couple went to the courthouse to get a marriage license, they were shocked by their reception.

They had to pass a lengthy line for divorces before they found the much shorter "Marriages" line. "You sign here," the clerk bellowed. "If you are illiterate, put a cross there . . ."

Documents? "You need none. You say, 'I want her,' she says, 'I want him.'"

Maria had hoped for more. "And what about the church?" she asked.

"If you are so stupid that you need it, you can have it," the clerk replied. Startled, Maria and Eugen hastily withdrew. On the street corner outside the courthouse, the new Mr. and Mrs. Jablonski shook hands and headed back to their jobs. Eight days later, they had a proper church wedding.

More complicated, however, was finding a place to live in a city still suffering from the aftershocks of war. As two intellectuals, they were eligible for an apartment with two rooms and a bath under the new regime. They were lucky to find one in the abandoned home of Eugen's rich uncle, who had fled the country. They shared this elegant house, its walls still adorned with tapestries, with tradesmen's families, who had the right to just one room apiece.[12]

Life in Bela Kun's Magyar Soviet Republic went from bad to worse. The communists terrorized the country, Maria remembered: "Danger everywhere. Night after night scores of people were taken from their beds—and dropped into the Danube." Food was scarce, and young men, such as Eugen, were particularly vulnerable. During the counter-revolution that summer, worried he would lose his job, Eugen fled to Poland, where he wrote "glowing letters" about life in his father's home-land. He found work as a surveyor for the new Polish government, though he was paid a pittance.

Despite the terror, Maria was reluctant to leave Budapest. At thirty, having survived and succeeded on her own for more than a dozen years, she was loath to give up her teaching job at the Commerce School, where she had become known for inventing a new, quick way of com-puting percentages. Few central European women of her day had earned such status. Ambitious and proud of her achievements, she talked over her dilemma with a friend. She recalled a time when Eugen was reading to her about Poland, and she suddenly did not hear his voice: "He was sobbing with his face buried in a pillow." She should go to him, her friend advised, but not let him know how hard it was to make the decision. "Well, I was not that big a person," Maria recalled. "But I promised myself something. I will never hinder my husband. I made his ambition mine, and I did everything in my power to make him succeed."[13] She soon left to join Eugen in Poland, but the bitterness re-mained. On a rare visit to Czechoslovakia during the Dubcek thaw in 1968, she told Wanda's cousin Vladimir that she had spent the happiest years of her life in Budapest. Abandoning her career was a huge loss, she had also told her daughter.[14]

As Maria had feared, life in postwar Poland was dreadful. What she found in the Carpathian Mountains horrified her: "Everywhere were burned down houses with nothing left but chimneys—graves of un-known soldiers—crosses in plowed fields. Orphans were so numerous that the fields seemed sown with them. . . . I just stopped to exist." But she could not. She had to help her husband, who was working his way through mountains still littered with the detritus of trench warfare. "My function," she later recalled, "was to overcome the million difficulties that life in a war-torn country put in my way." Although she did not speak Polish, she took responsibility for buying provisions. Virulent inflation plagued the region, so she had to fetch her husband's salary with a suitcase and then rush to the market for groceries. Prices mounted hourly. With salaries and prices in the millions, the scene was absurd: She saw a hundred-zloty bill trampled on the ground, unnoticed and untouched.[15]

After Wanda was born and his wife rejoined him in Poland, Eugen decided he had to find another job. His salary as a surveyor, his wife said, was less than the wages of his daughter's nursemaid. Maria was not happy living in a mountain cabin, and the Polish government was running out of funds. Private business looked more promising: A local company, Galacia Oil, hired him as a petroleum geologist. "My father could not afford me," Wanda later said, only half jokingly. "He got into oil because of me." Thanks to her father's improved prospects, she was able to join her parents in Poland when she was three or four.[16]

It was there that Wanda first encountered the smell and feel of petroleum. As she later enjoyed recounting, her only "claim to fame in terms of oil is that at the age of four, I got chased by an irritated goat into a pool of oil." She explained that in "those days they used to store crude oil in open pits, with little paths between them." To her dismay, "the goat and I ended up on the same path simultaneously: One of us had to give."[17] Although her mother cleaned her up, the oil stuck.

By 1925, Eugen's hard work on the geology of the Carpathian Mountains paid off with the publication of a book and a job offer from Vacuum Oil, a Standard Oil affiliate that became Socony-Vacuum and eventually Mobil Oil. Then the world's leading marketer of lubricants for machinery and cars, Vacuum needed to know whether oil might be found and produced in areas of potential new sales. Eugen could assure the company that in Poland, at least, they would find none. In 1926, at a conference at Mobil's New York headquarters, directors congratulated Jablonski for an astute presentation on Europe's oil potential and, not incidentally, for saving the company millions of dollars by discouraging them from drilling in Poland. In return, they offered him Texas.[18]

GLOBETROTTING

Eugen jumped at the chance to go to Texas. Botany and Poland, although close to his heart, held no future, whereas America opened up a world of possibilities. The 1920s were boom years for oil searchers; thanks to supply shortages during the war and a surge in demand for gasoline, prices for crude oil jumped 50 percent by 1920 to a lucrative $3 a barrel. Anxious to expand its production, Vacuum assigned Eugen to a desolate but promising region in West Texas. He arrived with his wife and daughter in 1927 just as the great Permian Basin, one of the world's largest oil fields, was discovered. "From that time onward," Maria later said, "we were always on the move."[19]

Texas proved an adventure and a challenge. Eugen and Maria quickly became Eugene and Mary, and under the tutelage of Vacuum's chief Texas geologist, Eugene painstakingly established his reputation. Technology was revolutionizing the process of geological surveying as new instruments, especially the seismograph, gave geologists new ways to see underground, but Eugene's instinctive feel for the land also served him well.[20]

In their first year in America, home for the Jablonski family was a series of bleak oil camps on the vast, barren plains of West Texas, as Wanda's father traveled from one possible exploration site to another. Wanda remembered this Gypsy-like existence as a lonely time. She found few, if any, playmates in the workers' shanties that the company had built on the treeless, sun-scorched red earth. She sought solace in the tumbleweeds, poking at snakes. The summers were oppressive, the winters frigid. Dust storms, fierce winds, extreme temperatures, and mosquito-infested tents made life harsh. Etched in her memory, Wanda later said, were two things about the Texas camps: the fearsome sight of oil fires and the dread of tarantulas.[21]

The arrival and departure of small propeller planes, which geologists used for aerial surveys, provided brief excitement. A rare family photo of that time captures the bleakness: Seven-year-old Wanda, a small solitary figure in a heavy wool coat and hat, looks out over an immense, empty landscape at a single-engine plane, kicking up dust behind it. In another, dated May 20, 1928, Wanda stands with her father

Wanda greeting a geologist's survey plane in barren West Texas in 1928.

and another couple in front of their desolate, windswept camp: a single row of a dozen wood-frame, tarpaulin-covered shanties.[22]

Wanda's mother, a cultured, erudite city girl, must have been in shock. But as Mary looked back on her first experiences in the United States some thirty years later, what she remembered was an extraordinary sense of freedom. Life in postwar Poland must have been unsettling, frustrating, even fearful for this self-declared Hungarian woman. What's more, family ties with her Slovak father and brothers were strained. Despite the drab loneliness of life in West Texas oil country, Mary was pleased to leave Europe. "I can't tell you," she later gushed, "what thrilled me most in America."[23]

Certainly she was awed with the vastness and beauty of the country. She and Wanda accompanied Eugene on several months-long trips through the West in a Model-T Ford with a trunk strapped to the back. In 1929 and 1930, they roamed from the beaches of Texas to the deserts of Arizona and New Mexico and delighted in examining desert flora: ocotillo, saguaro, cat's claw, and scrubby mesquite.

Undeterred by the deepening economic depression, they took to the road again in 1931, traveling from the Rio Grande to Canada and back, all "without a single frontier to cross," Mary reported with pleasure.

Wanda had two clear recollections of these travels. First, she earned a lot of coins from her father by identifying plants and flowers—a penny for every correct answer. Best of all, however, she learned to drive at age eleven. Since her mother feared driving and her father was easily bored during long hours on straight, often unpaved roads, Wanda was allowed to scoot in front of him and steer. "My father took the wheel when we passed through towns," she recalled. "He thought it would alarm people if they saw that the machine was being driven by a child."[24] Next, they headed south to Mexico, which had been the world's second-largest oil producer in the 1920s but was losing ground by the early 1930s as revolutionary turmoil drove away foreign investment. After two months, the family returned home, which by then was San Antonio.[25]

Wanda was late for school again. After a year of migratory camp life, the Jablonskis had stayed briefly in the grimy oil towns of Midland and Eastland before moving in September 1929 to the larger, more settled San Antonio, where she attended a Catholic school, St. Ann's. But in 1931, for instance, she did not start sixth grade until November. Eugene then left for nearly a year to work in Australia and New Zealand with Standard Oil of California geologists to appraise the oil potential. During his absence, Mary devoted herself to improving the English that she had picked up in small Texas towns and found "hard to correct." Her husband teased her, saying her real reason for taking night classes in English literature was to see a blackboard again.

Mary Jablonski was a tough teacher. Without her own classroom and often without her husband, she poured out her considerable energies on her only child. Devoted but exacting, Mary ensured that her daughter performed well academically despite all the missed school days. Wanda consistently got high marks, but Mary taught her more: unnerved by her devastating experiences after World War I and by the onset of the Great Depression, Mary impressed on her daughter the importance of thrift, a characteristic that marked Wanda for life. Though Wanda delighted as an adult in telling tales about her childhood travels, she talked little about her mother except to say that Mary had once whacked her so hard with a broom handle that it broke. A martinet, not a nurturer, Mary believed Wanda needed to be tough to survive in an uncertain world.[26]

The following year, 1932, brought another uprooting. After a lengthy visit to Slovakia and then five weeks in California, the Jablonskis traveled to New Zealand, where, on their first night, the country was hit by an earthquake. Mary found it exciting: "Mothers were holding crying babies in blankets, the water pipes were broken and water came through one ceiling to another; not one chimney remained; the church's rosette window fell out." The Jablonskis stood in the middle of the street, watches in hand, waiting for the next tremor—they counted one hundred that night. As the earth's crust cracked around them, Mary held Wanda's hand and told her, "Keep your nerve so that if we need to jump, we'll jump right."[27]

For two years the Jablonskis lived in New Zealand, where Eugene's exploration work sometimes proved embarrassing for his wife, who was often asked what her husband thought of New Zealand's oil potential. "Once, at a tea party," she recalled, "my hostess, a wonderful pianist, asked me the same question." To avoid revealing details of Eugene's work, Mary replied, "Of course, you know Paderewski, the great pianist, from his work." The hostess looked puzzled, since Ignacy Jan Paderewski had also recently been prime minister of Poland. "You know he had to resign because his wife talked too much." Startled, the hostess responded, "Oh we don't want that." Mary later called her experience in Australasia from 1932 to 1934 "one of the two nicest in my life." She and Wanda often accompanied Eugene to his survey locations, including the dramatic glaciers. They visited Admiral Richard Byrd's Antarctic expedition ship so that Eugene could inspect his surveying equipment. Wanda also got to know her father's friend, the eminent geologist Max Steineke, who would discover Saudi Arabia's massive oil fields. He became her hero, she later said, because he handed her—a rambunctious twelve-year-old—a rock-sampling hammer and invited her to join the experts.[28]

Wanda's parents were happy with her school in New Zealand, where she again generally excelled. "It was never necessary to ask Wanda to do her homework," her mother wrote. "More often I had to restrain her." However, her carefully saved report cards suggest a less than idyllic experience—and student. Arriving in midterm, Wanda "worked very well against a heavy handicap," one teacher wrote. But

Wanda rebelled a bit the next year. Her teachers noted she "has not worked this term" and "could be first [in class] with a little effort."

Still diminutive in size, Wanda began to assert herself in other ways. Though she had a Polish passport, she identified herself as an American. Five years of saluting the flag every day in school had made her an American at heart. This feeling intensified in New Zealand, where she insisted on using American spelling, even though it greatly annoyed her English teacher. She ate lunch "American style," with the fork in her right hand, even at the expense of ridicule. Although her mother found New Zealanders kindhearted and courteous, Wanda later caustically quipped that the sheep were more interesting than were the people.[29]

When her family returned to the United States in 1934, Wanda bought an American flag with her own money and slept beneath it. Nonplussed by their daughter's willfulness and determination to stay there, the Jablonskis faced a dilemma. Socony-Vacuum had reassigned Eugene to Europe, and Wanda adamantly refused to go. Although her parents conversed in Hungarian and English, Wanda spoke only English fluently, so she had been miserable during her family's visit to Slovakia in 1932. She also felt uncomfortable with her Slovak relatives because they knew so little about her life. Her grandfather had once interrupted Wanda's earnest attempt to describe her travels in America with the question, "What is Texas?" Acquiescing to Wanda's desire for American roots, her parents chose a Catholic boarding school for girls in New Jersey, the Academy of St. Elizabeth. She attended only one year. Wanda, then fourteen, "was not contented there," her mother said, so the summer of 1935 she rejoined her parents in Slovakia.[30]

That autumn, Wanda's itinerant father took the family to Egypt. Socony-Vacuum, though still a bit player on the oil-exploration scene in 1935, wanted to expand its foreign supply sources beyond its meager share in the Iraq Petroleum Co. It was constrained by the infamous 1928 Red Line Agreement between the major oil companies, which limited exploration in much of the Middle East by precluding the signatories— British, French, and American—from operating independently of one another within the old Ottoman Empire boundaries. With encouragement from their governments, the companies thus sought to control the pace of oil development in the face of a mounting worldwide glut. Socony-Vacuum set its sights on North Africa, a promising but largely unknown region.[31]

Cairo, long fancied by the British and French to be the most European city in the region, proved an attractive base for the Jablonskis. Mary settled into an English *pension* with a beautiful view of the Nile, and her daughter attended the French school École Merion. Wanda boarded during the week at the home of the schoolmaster because her mother thought she should acquire a "correct" accent and insight into French family life—an arrangement that freed Mary up to take archaeology classes. This situation must have added to the emotional distance already present between mother and daughter.

Still, Mary was determined to expose Wanda to as much adventure and culture as possible. With a company car at her disposal, Mary took Wanda on weekends for archaeological expeditions and swims in the Suez Canal. A favorite site was the Sphinx, where, one night of a full moon, they sat on the statue's paws. Although Mary was enthralled with Egypt, Wanda, a moody teenager, saw it as yet another foreign environment to which she had to adjust. Wherever they toured, children would press up against them begging for money. Family photos show an uncomfortable-looking Wanda in a prim dress and high heels astride a camel; another picture shows Egyptian women and children staring at her. As she later said, "I learned adaptability."[32]

Delighted to be free of her New Jersey convent school, Wanda also found Egypt a place to test her limits. She later told a friend, British Petroleum Chairman Peter Walters, about a time when she and several girlfriends got drunk. One passed out, and the others, "thinking she was dead, rolled her under a bunk, and didn't know what to do. Fortunately, the girl recovered." Wanda recounted to several friends that she was so rebellious, she once escaped from school and hid in the Arab *souk*.[33] She also delighted in talking about her first major venture into the Egyptian desert. As the story goes, her father took her, without her mother, on an extensive camel trek, possibly toward Libya. For Wanda, it was rough going. The camel bumped her behind, she wailed, and he burped and gave her lice. She complained of boredom after days and days on the endless sands. Finally, Eugene's patience with his fifteen-

year-old ran out, and he sent her packing, without a word of explanation. He headed west, while she was taken east by some of his Bedouin
helpers. Sometime later, she ended up on the outskirts of Cairo, where
her father met her. She learned, she later said without a trace of rancor,
never to cross him again. And she never forgot the delousing.[34]

Eugene was a driven man. Having turned his passion for botany into
oil prospecting, he was, above all, happiest in the field, "exploring off
the road and in backward places," his wife said, than sitting behind a
desk. Somewhat hard of hearing, he enjoyed solitude: "He would disappear on long trips, studying botany, geology, and ethnography," his
daughter recalled. Inattentive to family finances and unconcerned
about where he would eat his next meal, he relied on his wife, she
noted, to keep "the home fires burning." Although inside "he was made
of steel," his future son-in-law, Jack Jaqua, remembered, Eugene was,
nonetheless, a gentle man. He had a "hell of a good sense of humor
and liked to quote Shakespeare," Arthur Mills, Mobil's London office
chauffeur recalled. (Mills would later drive Wanda when she visited
London in the 1950s and 1960s.) Other friends said he enjoyed a good
party, and especially loved to dance to Polish folk music. Though absent
for much of Wanda's childhood, Eugene imparted to her an intensity
and a single-minded professionalism. He also encouraged a sense of adventure and fun.[35] Given the stories she told of her youth, recalled
Sheikh Yamani, he thought Wanda was always trying to prove to her father that she was as good as any boy.[36]

Despite Wanda's troubles on her first desert venture, she soon joined
both parents on another major expedition: a car trip to Palestine, crossing the Suez Canal and the Sinai and Negev deserts. Eugene brought
along several colleagues. Although Mary was pleased to see Mount
Sinai, she came to agree with her daughter that traveling through the
desert with geologists was tedious: "I never saw so many drab stones in
all my life. As far as one's eye could reach, nothing but stones — not even
a blade of grass." After fifteen nights of uncomfortable camping under
tarpaulins, she had her reward: Christmas night in Bethlehem. It was,
she said, the "nicest Christmas present" from her husband.[37]

When Eugene took off on another expedition, this time to Sicily,
Wanda finished her school year in Cairo but was soon on the move

*Wanda with her father and a geologist at a temporary camp in the Sinai desert,
December 1935.*

again. Mary, the inveterate teacher, sought to round out her education
in classical history by taking her on more trips: they headed up the Nile
to Luxor and then traveled through Greece and Italy on their way back
to Slovakia for the summer. Along with some cousins, Wanda attended
a scout camp near Banská Bystrica run by her uncle, Ladislav; she
learned how to saw wood, tie knots, and sing Slovak songs, despite her
limited knowledge of the language. But Slovakia proved more interesting when she was sixteen than it had when she was twelve. Though only
really conversant in English and French, she quickly got to know an appealing young man, Stefan Gregorovic, a cousin's friend who taught
French at a nearby high school. They met several times at the scout
camp and then again briefly on a visit to Vienna. The attraction was mutual. Three years later, in the only surviving letter of their correspondence, Stefan wrote that their days together "are the moments in my life
to which I attach not only my most fond memories but also a deep feeling of understanding and affection that has grown and still persists."[38]
    Wanda did not stay in Slovakia long enough to know Stefan well.
Once again, she changed schools—at least the sixth time in ten years. "I
was always taking placement examinations," Wanda remembered.
"They didn't know what to do with me." This time her mother, certain

Wanda at about sixteen, a student at St. George's School, Hertfordshire, England.

of her daughter's "brilliant mind," took her to England and enrolled Wanda at St. George's School in Hertfordshire for her last two years of high school. Wanda recalled that her grandfather approved because he mistakenly thought that the Church of England school, named for the country's patron saint, was Roman Catholic. No one disabused him of that notion. An elegant boarding school with spacious, manicured lawns, Victorian Gothic architecture, and wealthy students, St. George's proved to be another source of culture shock to this itinerant young woman: "Because of my different attitude toward the world, it was very hard for me to fit into a small English or French society." Nonetheless, she adapted and succeeded.[39]

By 1937, Eugene had become Socony-Vacuum's chief geologist for Europe, based in Hamburg—at least for eight months. Wanda came for holidays and spent the summer studying music and German. Brandon Grove Jr., the son of one of Eugene's colleagues, remembered the Jablonski apartment in Hamburg as a warm, lively place. Grove's mother, an expatriate Pole, pronounced their name the Polish way, "Yaboyski." The Groves' son, later a U.S. ambassador, recalled the "party feeling" at the Jablonskis': Mary, with her penchant for colorful central European shawls and deep appreciation for classical music, and the "kindly, fun-loving" Eugene, with his love of folk dancing. "Europe was going to hell, but dancing for the pure joy of it was a way for these expatriates to express their ethnic roots." The Groves' son distinctly remembered not liking Wanda. At that time, he was nine years old, and Wanda was about seventeen. Once, she was told to "go play with Brandon," which she made very clear was "the last thing she wanted to do." His parents also held up the brilliant Wanda as a model to him: "See how much Wanda reads," he was told. "I thought her nasty."[40]

Wanda, for her part, recounted only one provocative anecdote from those days—her encounter with Nazi Field Marshal Hermann Göring. Eugene, apparently thanks to his position at Socony-Vacuum, secured a private box for a performance in 1938 of one of Wagner's operas and took Wanda along. Many Nazis attended, and Göring was seated in a nearby box. Amid all the pomp and circumstance, Wanda laughed heartily because she thought it "so funny to see [Göring] sitting with his fat bottom on a little gold stool."[41]

But for her family in Slovakia, the threat from Germany was no laughing matter. Older members of Father Hlinka's party, wary of Hitler's policies toward Catholics, had misgivings about his National Socialist ideology. Younger militant nationalists, however, formed the Hlinka Guards, a paramilitary fascist group that rapidly gained strength. Wanda's friend Stefan joined the guards, convinced that the greatest danger to Slovakia and Catholicism came from Czech leaders collaborating with Russian Bolsheviks. When Wanda wrote to him that she hated uniforms, he responded, "The uniform that I wear is that of peace, order and defense." He was thrilled that Hitler had rescued Slovakia from the "dangers" of communism. Even Wanda's grandfather, though wary of Germany, thought Hitler preferable to "godless communism."[42]

UNCERTAIN CITIZEN

In the midst of these troubles, Wanda faced tough choices about her education. In her two years at St. George's, she had excelled in Latin, German, French, mathematics, and music, and in June 1938 she took the grueling matriculation exams for Oxford and the University of London. Of the hundred students who passed the London exams, Wanda was one of seven women. Oxford admitted her, but because of her age—she did not turn eighteen until August 20—she had to wait until the following year to attend. Since her father had returned to New York that spring to become chief geologist at Socony-Vacuum's New York headquarters, Wanda and her mother followed in midsummer. Convinced that she could attend one of America's best universities for a year before enrolling at Oxford, Wanda wrote to Harvard, Yale, and Princeton. Dumbfounded to learn that they did not admit women, she decided to enroll at another Ivy League school that did—Cornell. With the onset of war in 1939, she had to abandon her plans for Oxford and got her bachelor's degree from Cornell.[43]

Wanda graduated near the top of her class, but in her first year she was forced to abandon her plan to become a physician. In biology class, she could not bring herself to dissect a frog. She tried first in the lab, but then took the frog to her dorm room where she thought she would have the courage to carve it up in private. But "I could not cut the frog!" Wanda exclaimed.[44] She switched to international affairs.

In art history class, Wanda was seated alphabetically next to her future husband, Jack Jaqua, who remembered Wanda as "very bright, alert, eager for action." Two years her senior and managing editor of the Cornell newspaper, the *Daily Sun*, Jaqua found Wanda "interesting," but he was dating another woman. Wanda, too, had her share of beaus—family photos show her in evening gowns on different occasions with different men.[45] By 1939 she had already turned down two marriage proposals. Indeed, at that point she began to express her affection for Stefan Gregorovic more explicitly, prompting a lengthy love letter from him. She asked him to visit her in the United States. Stefan's reply— one of the few personal letters she kept for a lifetime—disappointed her. He wrote of his affection, but, since he was conscripted, he could not leave the country. Would she come to him? Much to her dismay, Wanda soon discovered that she could not, even if she wanted to: she was not an American citizen.[46]

When Eugene became a naturalized American in 1938, Wanda assumed that, as his child, she had automatically become one, too. But two years later, in an international-relations course, she realized that might not be true. She had entered the country with a Polish passport, but she was born in Slovakia, now an Axis state. She wrote to explain the situation to the Department of Justice, but by the time the government ruled on her case in 1943, she was attending Columbia University's graduate program in public law and government. She was stunned by the news. How could her parents have failed to take care of this? Years later, Wanda could still remember the letter's exact phrasing: "According to the alleged facts in your case . . ." The department, in a routine wartime decision, had determined that she was an "unfriendly alien" because of her birthplace and confined her to the island of Manhattan.[47]

This rejection by her adopted homeland and the egregious parental mistake that allowed it left an indelible mark on Wanda. Ever since her childhood in the West Texas oil fields, she had considered herself American. Jaqua remembered her speaking with a "slightly exotic accent" at

Cornell, but she soon got rid of it. As a graduate student at Columbia, with the outcome of the war still uncertain, Wanda found herself stateless. Much later, she would say that New York was the only place in the world where she could live, the only place where she eventually felt comfortable; but at that time, her confinement to Manhattan and her "unfriendly alien" status was shocking and intensely embarrassing.[48]

The denial of citizenship, thanks to her parents' oversight, compounded Wanda's sense of personal uncertainty. Her itinerant life had bred a certain toughness and adaptability, but also a sense of not belonging, of being the outsider wherever she went. As an only child, she developed the assuredness of one accustomed to an adult world, but also the emotional distance of one not used to playing in a group. She had no fear of asking questions or speaking her mind, but behind that bravado lay a shyness, a craving for acceptance and affection, an anxiety about the future.

Later in life, Wanda told friends that she felt somewhat abandoned as a child. She believed she was secondary to her parents' marriage, her father's work, and her mother's self-absorption. Eugene and Mary, both ambitious and driven, fostered these tendencies in their daughter, but Wanda's relationship with her father was better than that with her mother. Despite his frequent absences, they had enjoyed many travels together and she shared his infectious curiosity and sense of fun, as well as his love of dancing. From her mother she learned thrift, hard work, and considerable sophistication in traveling from one continent to another; but Mary, while she was devoted to her daughter, lacked maternal instincts. Wanda remembered her beatings, not her care. A rather cold and bitter person, Mary was marked by her own series of losses: the deaths of her mother and sisters and the end of her promising teaching career, followed by financial insecurity and an uncertain national identity. She particularly resented having to abandon her professional life and depend completely on her husband and his work.

Wanda, like her mother, had a survivor mentality. Her nontraditional upbringing led to a fierce independence and drive to excel. However, repeated losses of home and friends, as well as the physical and emotional distance from her parents, left emotional scars that Wanda struggled with all her life.[49]

In August 1943, shortly after she found out that she had been denied U.S. citizenship, Wanda quit her graduate studies just short of completing a master's degree with a specialization in international affairs. At Columbia, she had lived at the university's International House, taken courses on the Middle East, and attended a journalism class. As Americans struggled with a grueling war, Wanda badly wanted to contribute in some way, but her alien classification prevented her from joining a women's auxiliary corps. By then she was corresponding regularly with Jack Jaqua, who, after two years in law school, had joined the Marines and was fighting in the Pacific. Many of her college and graduate-school friends were also in uniform. Given this atmosphere, graduate studies may well have seemed tedious. After the 1943 summer session, Wanda left Columbia without turning in her thesis to complete the degree. Her only published comment on her international affairs program was "God, was that dull!"[50]

She sought work at the Council on Foreign Relations, but was turned down because she could not type. Desperate to find a job, she became a waitress, but lasted only one day: her arms were so weak that a tray wobbled and fell. Crushed again, she did not give up. Her parents had financed her studies, but she was determined to find a way to support herself. When an employment agency sent her to apply for a job as a messenger at the *New York Journal of Commerce*, she jumped at the chance.[51]

2.

"Just Call Me Bill"

Weak arms and a lack of typing skills were not handicaps for newspaper messengers. They just had to be quick on their feet. Also known as copyboys, these messengers took care of the paper flow in a big newsroom. Leaping up at the cry of "Copy!" or "Boy!" they grabbed the latest stories from reporters and delivered them to editors, tore off wire reports from clattering teletype machines, and fetched pictures from the photo department. They kept the staff pumped with coffee, sharpened pencils, and cleaned paste pots.[1]

In late 1943, Wanda convinced her interviewer at the *Journal of Commerce* that she had the necessary attributes to be an excellent copyboy: strong legs and lots of energy. She also answered one critical question correctly: "When I said I had no claims to be a writer," she later recalled, her prospective employer "said that was good, because he did not want someone on his staff aspiring to write the 'great American novel' on the *Journal*'s time." Her timing was also good—being female was not the almost insurmountable barrier it had been only a few years earlier. The war had created such a shortage of labor that many companies, including newspapers, were forced to hire women in nontraditional jobs. But this twenty-three-year-old did not last long as a copyboy. Within months, she landed her first front-page story.[2]

BUSINESS NEWS

Business reporting was, in the years right after World War II, one of the least developed specialties in journalism. Of the three main categories of the written press—newspapers, business publications (dailies and magazines), and the specialized trade press—all provided limited coverage of business developments. At the metropolitan dailies, politics, foreign news, sports, the arts, and even the police beat were all more interesting avenues for reporters. Business sections were notorious for publishing rewritten press releases. Many business reporters, said one who began in the 1940s, were "often tired rejects, relegated to the business page because the managing editor couldn't think of anywhere else to send them." What's more, given minimal disclosure rules, executives from publicly traded companies often refused to reveal basic facts.[3]

The business press was not much better than the general dailies for providing business news. Among the leading magazines, *Fortune* alone had a reputation for in-depth, substantive stories. Not until much later did *Business Week* evolve, according to one critic, "from a dry, bland, unvarnished cheerleader of business and businessmen into a tough, probing, perceptive analyst." The *Wall Street Journal*, originally a financial paper with limited circulation, only began in the 1950s to publish more wide-ranging national and international articles. Its main rival was the *Journal of Commerce*, founded in 1827 and noteworthy as the oldest continuously published newspaper in the country. Stodgy but well respected, with a long-standing commitment to free market enterprise, the *Journal of Commerce* held its ground in those postwar years as the paper of record for New York commerce, specializing in shipping news, commodity coverage, and foreign trade reports.[4]

DRESSED OR UNDRESSED CHICKENS?

Wanda was still technically a copyboy when the *Journal of Commerce* published her first byline story on March 1, 1944. The paper happened to be short-staffed that week, so Wanda got an unexpected summons. Her first subject was coal: in rather long-winded prose, she reported that the coal industry was upset with government interference in the market. "Though the emergency nature of this action is appreciated in the coal industry," she wrote, using the passive voice for nearly every attribution, "apprehension is felt that this may be used as a precedent for future Government stockpiling and marketing of coal." Despite the tentative start, she addressed a major concern of the *Journal of Commerce*: the appropriate role of government, given the vast array of federal wartime controls. Wanda's perceptiveness caught the eye of a key editor.[5]

Heinz Luedicke, the managing editor, was remarkably accepting of female reporters. This German émigré, who set the pulse of the paper, had several women on a staff of about fifty. "Luedicke's attitude toward women was totally unusual for the time," said Eileen Shanahan, a *Journal of Commerce* reporter in the late 1950s. "I never felt put down by him in the slightest." For Wanda, he became a mentor. "He didn't care if you were a man or a woman as long as you got the job done well," Wanda's future husband, Jack Jaqua, later recalled. "He really encouraged Wanda." Under his tutelage, she soon became known as *Wunderkind*—the amazing child.[6]

In those days, the newspaper employed a number of colorful characters. "It was a great place to work," Eleanor Schwartz, then a young assistant, remembered. "It was very noisy and [had] a lot of eccentric people." On the third floor of 63 Park Row in lower Manhattan, the newsroom was a cavernous space crammed full of old oak desks and Underwood typewriters, telephones, and teletype machines. From a glass-enclosed office, "Dr. Luedicke"—the staff never called him "Heinz"—presided, Schwartz said, while his flamboyant secretary, who delighted in wearing black bras and revealing blouses, openly talked about her affair with her boss. Nearby, T. K. ("Art") Kramer ruled the copy desk with booming voice and dangling cigar. "Kramer was always yelling at me because I was always filing on deadline," Shanahan recalled. "But he was a very competent editor, who really got the paper out every day."[7]

Despite his short temper and irascible tongue, Kramer met his match in Hazel Stanton, his chief assistant and headline writer. A descendant of the feminist leader Elizabeth Cady Stanton, Hazel Stanton defied all norms of polite womanly behavior. "She would pick up the phone and say hello in such an elegant, educated voice," Schwartz recalled. "As soon as she heard something she didn't like, she'd let loose this stream of language that would make a truck driver blush." Nearby sat "chemicals" editor Dwight Moody, who had two-foot-high piles of paper teetering precariously on his desk. "But if you asked him the price of, say, yling-yling oil," said Schwartz, "he'd fish out a sheet from the bottom of a pile and say, 'Here it is!'" The atmosphere was so informal that the receptionist, Jaqua remembered, "sounded as though he had just come in off the range." Poker games were common. In this relaxed, collegial setting, reporters had considerable latitude and were subject to minimal editing.[8]

It did not take Wanda long to establish herself as a credible reporter. Buoyed by her initial success with the coal article, she volunteered to cover stories that no one else wanted, particularly trade-association affairs. "My boss just told me, 'Write down what you hear,'" she later said, and in general the strategy worked. Not only did she get in print, but she got free meals. Her pay was so low—starting at $16 a week—that she had to take on anything she was assigned, including the peanut, linseed, and castor-oil markets. At a luncheon for poultry purveyors, however, "her first question 'What's the difference between dressed and undressed chickens?' proved her exit line," the *Washington Post* later reported in a 1957 profile of Wanda—for the paper's women's page.[9]

She was soon digging up enough original stories to get several front-page bylines a month. "She hit the bull's eye," the *Washington Post* noted, when she wrote an original piece analyzing the likelihood of fierce competition between coal and oil. When the *Journal of Commerce*'s oil reporter left shortly thereafter, Wanda "nipped into his slot on a temporary basis," and she was not temporary for long. In late November, she landed a page 1 scoop about emergency military needs causing an unexpected fuel oil shortage, and the merits of dressed and undressed poultry were history. Wanda was in her element: derricks and yields, sweet and sour crude—this was the language she had learned as a child. Three days later, she was back on the front page with a story about an oil discovery in England that was kept secret for military reasons. Geologists were skeptical about the significance of this find, she reported, but added wryly, "The east Texas area was abandoned as worthless two years before the great oil discovery was made there."[10] She knew about this world, thanks to her father. She could put the news in context.

Press coverage of the U.S. oil industry at that time consisted of occasional stories in metropolitan dailies and business magazines, parochial reports in the Texas and Oklahoma daily newspapers, and extensive but generally docile reports in the trade press. Oklahoma's *Oil & Gas Journal*, the country's most profitable magazine thanks to hundreds of pages of advertising, touted itself as "the authoritative spokesman" for the industry, but its tone was reverent, not analytical. Other major sources of oil news—the weekly *National Petroleum News* and the daily *Platt's Oilgram News* and *Platt's Price Report*—gave regular, detailed coverage but were not known for analysis or exposés.[11]

In March 1945, a few months after her first front-page stories, Wanda scored another coup that drew her right into a major reassessment of Washington policy. She helped put together the *Journal of Commerce's* first annual oil-industry "special," eight pages on the postwar outlook with articles from company presidents and other luminaries. In it, she tackled head-on the two most divisive issues roiling the American petroleum industry: Was the country running out of oil, and should the government actively promote development of foreign supplies? International oil companies based in the United States predicted that the country would soon shift from being a net exporter to a net importer of crude. Domestic producers disagreed, inveighing against the "dour prophesies" of "alarmists" who claimed the country was on the brink of "oil starvation."[12]

The lead commentary went to the secretary of the interior at the time, Harold Ickes, who insisted that the nation needed access to foreign reserves, especially Saudi Arabia's prolific fields: "War has shown us that there can be no isolationism in oil." Ickes did not reveal that, behind the scenes, he was trying to engineer a policy shift toward greater support for U.S. exploration abroad, but a clear hint at his strategy appeared in an article alongside his—by Wanda Jablonski. According to "well-informed" sources, she reported, the industry would get "much more vigorous Government backing" in their efforts to secure foreign supplies, "an abrupt departure" from previous policy, because

Franklin D. Roosevelt's administration recognized oil's geopolitical importance. By reading the signals correctly, Wanda was more explicit about Ickes's shift in strategy than he was willing to be in print.[13]

KING OF THE HILL

This special issue on petroleum established Wanda's credibility as a reporter. In less than four months at the oil desk in 1945, she had secured her job. Shortly thereafter, on May 11, she also found a measure of security in her personal life: she married Captain John Clayton Jaqua Jr. and, thanks to her American husband, obtained American citizenship.

Jaqua was a native of Winchester, Indiana, who was known at Cornell as "King of the Hill." Small, lithe, and dark-haired, he was an "exceptionally attractive man," according to Wanda's *Journal of Commerce* colleague Eleanor Schwartz. "He looked like the movie star John Lund." He was also a man of considerable intellectual gifts. After graduating from Cornell in 1940, he attended Yale Law School, where he was editor-in-chief of the *Yale Law Journal*.[14] During the early years of the war, Jaqua recalled, he and Wanda had had "some contact, but it was not serious" until he enlisted in the Marines. Shipped off in 1943 for the Pacific, he fought with the Third Marine Division on Guam and Iwo Jima. The two wrote frequently and got engaged by correspondence. In a brief newspaper notice on May 22, 1944, Dr. and Mrs. Eugene Jablonski announced the engagement, but it was not until early 1945—probably March or April—that Captain Jaqua returned to New York. After two years of active duty, he got a brief rotation at the Brooklyn Naval Yard while Wanda lived in a studio apartment in Manhattan's Greenwich Village. "We saw each other a lot," he recalled, "and decided to get married in a hurry" because he expected to return to the Pacific soon for the final assault on Japan. War had a way of bringing people to quick decisions.[15]

Jaqua remembered Wanda as a "very enthusiastic person, who loved to be at the center of everything." Wanda was great fun at parties, and she had a quick wit and an "ebullient personality." She would often regale friends with stories about her reporting experiences, itinerant childhood, and unusual travels. "People really liked her," Jaqua said.

Wanda with Jack Jaqua, her husband, in the late 1940s.

"She was a master of anecdotes." Petite and fine-boned, she was attractive, Jaqua recalled, but not a distinctive dresser, favoring tailored career clothes. Schwartz, a lifelong friend, was less flattering: "Wanda, in those days, was mousy. She never wore anything but gabardine suits with little white collars. She was not what you would call well dressed." She also had a prominent nose, which three decades later she had surgically reduced. What struck Jaqua the most, however, were her hazel eyes: "She had large, very expressive eyes, that held you with such intensity."[16]

That intensity carried their marriage for a number of years. Wanda worked long hours at the *Journal of Commerce*, rushing from one assignment to the next. She often worked late nights and weekends. "I remember how you would come dashing in with your shoulder bag stuffed full," an oil company secretary later wrote to her. "In those days it wasn't easy for a young woman to compete with the men in the business world, was it?"

Her ambition was obvious. "She was hungry for success," her new husband recalled. He too, had high aspirations and a promising future at one of Wall Street's preeminent law firms, Sullivan & Cromwell. "We were very wrapped up in our own work," he said, but still had fun together. "Often I'd call her about 9 P.M. and we'd arrange to meet on the subway. We loved going home together." They lived frugally in her tiny apartment with little more than a kitchenette and a fold-out bed. Wanda's lack of interest in cooking or children mattered little then. Nor did her decision to keep her maiden name professionally.[17]

SETTING HER OWN STYLE

Despite her rapid rise from copyboy to oil reporter, Wanda was nervous about conducting interviews. "She was determined to overcome her shyness," Jaqua said, but her friend Eleanor Schwartz remembered that it "took her a while to hit her stride. She had a lot of insecurities, but while she was reticent in some ways, she went after her sources with zeal." When Wanda needed to make a long-distance call to Phillips Petroleum, for instance, she was so concerned about the cost that it took her a while to work up the courage to ask for approval.[18] But she learned to turn her inexperience to her advantage. In thank-you letters to people she interviewed, she readily admitted her novice status. "I am painfully aware of my inadequacy in dealing with such complicated subjects after so brief a visit," she wrote in a typical letter, welcoming any corrections to her stories. This approach worked. "She was great at knowing how to ingratiate herself," Jaqua remembered. "She played the little girl with the big men. But she also built a very good reputation very quickly as someone who really knew what was going on, who really knew the market." Rather than mimic the cocksure tone of many reporters, she would coax her sources into giving her information—often far more than they intended—because, she explained, she "just didn't understand."[19]

Wanda also figured out how to circumvent public relations personnel by establishing a personal rapport with company presidents. She would corner them at the end of press conferences, pepper them with additional questions, ask for private interviews, and then send them clippings of her articles. Whereas many on the *Journal of Commerce* staff were "just commodity market reporters who followed stodgy markets, like burlap, and didn't do that much legwork," Schwartz said, Wanda became known for her investigative work. "She just loved," Jaqua said, "to go for the scoop." Indeed, she lived for it.[20]

Wanda tackled the New York fuel oil market with gusto. Although the paper published commodity market prices, its reporting of local oil prices, used for many city contracts, had not been dependable. One of her contacts, a fuel oil dealer who later became a company chairman, recalled that it took Wanda about a year to figure out who was reliable: "Buyers and sellers were trying to influence her, so it was no easy task. Some of them used her to sway the market but she learned quickly." Sometimes, when the food was bad at industry parties, he and "some of the boys" would take her to dinner: "I remember thinking, 'How could a woman really break into this all-male world of oil?'" But she developed a reputation for integrity so quickly that many more contracts were tied to her price lists. "She couldn't be bullied," Schwartz said, "or budged." What's more, while checking for prices and contract terms, she would sniff out news leads. Once she caught a whiff of something, she would backtrack, calling all of her sources again to get confirmation. A classic newshound, Jaqua said, "she loved the chase." Another *Journal of Commerce* colleague, Aaron Riches, said that she "almost single-handedly put the *Journal*'s oil page on the map."[21]

Over the next few years, Wanda expanded her coverage, rotating daily statistical tables with prices for all kinds of refined oil—gasoline, kerosene, diesel and residual fuel oil, "red oils" and "pale oils," plus "posted tank wagon prices." She began to beat the competition regularly: She was first to learn, for instance, about a change in plans for two huge East Coast pipelines; first to reveal government support for American interest in developing China's oil resources; and first to report Iraq's plan to build its own refinery, undermining Britain's monopoly. She also showed an eye for spotting long-term issues and trends, such as a report on American oil exports for civilian use in Europe. By 1946, her analytical pieces regularly appeared on page 1, and in late 1947 she won approval to start her own weekly column, "Petroleum Comments."[22]

A RESPECTABLE WOMAN

During World War II, more women than ever before ventured into American newsrooms, but with limited success. Some dailies and news services banned women from their domestic staffs, while hiring a few for their foreign desks; others insisted on the reverse policy. Several dozen women managed to become war correspondents, but they faced enormous discrimination. Until late in the war, the military banned women from front-line visits and even from most press briefings. When servicemen returned at the end of the war to reclaim their jobs, few female reporters kept theirs. United Press, for instance, fired all but three of its women. Even Eleanor Roosevelt, who had boosted women's prospects in the 1930s by restricting her press conferences to female reporters, could do nothing to stem the tide. "Women would not return to the nation's newsrooms in any numbers again until the 1960s," according to historian Kay Mills. Only the most dedicated women of that wartime generation, reporter Nan Robertson writes, "those who had lost their hearts to journalism and were persistent as well as talented, stayed on in newspaper work."[23]

Wanda Jablonski, both persistent and talented, never considered leaving. But neither did she challenge the rules of the game—at least not then. Women at the *Journal of Commerce* were paid less than men, but she insisted later that the disparity did not bother her, and she never asked for a raise. She loved her job. "I never would have gotten hired if they had had these female rights then," she said in 1976. "I was hired because I *could* be underpaid. The most important thing is opportunity, not equal pay." She believed in equal pay for equal work, but was grateful just to be hired and always worried that she might lose her job to a man. She did not question another unwritten *Journal of Commerce* rule of the 1940s: except for one older reporter, Emma Doran, women kept their sex hidden by using initials in their bylines. For the first decade of her career, Wanda received many letters addressed to "William."[24]

On a reporting trip in 1945 to western Pennsylvania, birthplace of the U.S. oil industry and a region with a significant but declining rate of production, Wanda arrived at the appointed train station, expecting to be greeted by several local executives. When the train pulled out, she found herself standing all alone on the platform, except for several men looking for someone they could not find. She picked up her bags, walked up to them and, to their astonishment, introduced herself as "W. M. Jablonski from the *Journal of Commerce.*" They had been looking for William, not Wanda.[25]

In another instance of mistaken identity, a Pennsylvania magazine, the *Oil Light*, reprinted in 1950 a tongue-in-cheek column by Wanda that poked fun at the industry's ineptitude in dealing with unpredictable weather. "While *Oil Light*'s policy is to avoid wherever possible reproducing articles," the editor wrote as an introduction, "the following very timely story by William Jablonski, Petroleum Editor, *New York Journal of Commerce*, was too good to pass up." Wanda thanked the editor for sending her the "very kind reprint," adding, "In the future, though, please just call me 'Bill.' With best regards, Wanda Mary Jablonski, Petroleum Editor."[26]

Newsrooms remained male provinces with male imprints—filthy floors, cigar smoke, beer bottles, and poker chips. Reporters "wore their press passes in the bands of their snap-brimmed fedoras, barked into stand-up telephones, and spat into their cuspidors," writes Nan Robertson, a former *New York Times* reporter and Pulitzer Prize winner who worked for years on the women's pages. By the 1950s, women who had been welcomed at the city desk in wartime "couldn't get past the front desk," notes Kay Mills. Women were not supposed to be subjected to the danger and degradation of "hard" news, recalled Marylin Bender, who eventually became a *New York Times* business reporter but only after covering fashion for years. Female reporters, particularly married women, were treated as pariahs, both by men and by noncareer women.[27]

When women did manage to write on business or economic matters, they had to struggle against crippling stereotypes. Wanda got her break thanks to Luedicke, but Eileen Shanahan, who began her career during the war and became the *New York Times'* first female economics reporter in the 1960s, was stymied in her efforts to cover business news. When she interviewed in the mid-1950s with Willard Kiplinger, founder of a prestigious business newsletter, he had no interest in her except "to see what kind of woman would apply for a job doing the kind of reporting *The Kiplinger Letter* required—inside information." After all, he told her, "A respectable woman, the only kind of woman we would want here, just couldn't do it." When she tried the *Washington Post*, she found that the assistant managing editor, Al Friendly, also subscribed to the stereotype: he liked her work, but "an economic story under a woman's byline," he said, "just wouldn't have any credibility." The receptive editor who finally gave Shanahan a job was Wanda's champion at the *Journal of Commerce*, Heinz Luedicke.[28] When she eventually moved to the *New York Times*, she had to "prove every day," she said, "that I was sane." Her few female colleagues encountered similar handicaps. Robertson, hired in 1955 as a fashion writer, took a long time to gain confidence because she was "affectionately" patronized. "Like many women of my generation," she writes, "I was not taken seriously."[29]

What's more, a double standard prevailed. Typical assertiveness in a male reporter was generally seen as unappealing aggression in a female. "So many people," Shanahan later recalled, "find tough-mindedness in a woman an unattractive trait—even now." When she once asked a pointed question at a press briefing with President Eisenhower's treasury secretary, the former steel executive ignored her. Finally a male reporter said, "Mr. Secretary, Miss Shanahan has asked a perfectly appropriate question. Would you please answer it?" In 1964, Shanahan met the *Wall Street Journal*'s Washington bureau chief, who told her, "If you were a man, I'd hire you." She replied, "I don't remember applying."

Wanda also knew how to respond in kind. Once at an industry meeting, a man, assuming she was a secretary, asked her to take notes. "Oh, I'd be happy to," Wanda replied, "as long as you type them up afterwards."[30]

Discrimination against women reporters was pervasive. Washington's National Press Club, a prized podium for presidents and prime ministers alike, not only banned women as members but forbade them from setting foot inside the club. Women protested to no avail until 1955 when the club allowed them to attend major speeches, but only if they entered through the back door and climbed the service stairs to the ballroom's balcony. "It was humiliating," Robertson writes. Women could see and smell the food—and the action—but could not partake. Often standing-room-only, especially with television cameras, the small balcony was hot and stuffy, the speeches hard to hear. Women reporters could not pose questions of speakers, seek comments, or get access to telephones—the basics of news coverage. Not until 1969 did the club finally vote to allow women onto the ballroom floor, and not until 1971 did it admit women as members.[31]

That a woman reporter would venture into writing about the oil industry was highly unusual—with one outstanding exception. In 1902, *McClure's Magazine* published Ida Tarbell's extraordinary series on John D. Rockefeller, which contributed directly to the dismemberment of his vast empire, the Standard Oil Co. Tarbell, as a Rockefeller biographer notes, turned America's most private man into its most public— and hated—figure. Reprinted in 1904 as *The History of the Standard Oil Company*, Ida Tarbell's indictment of America's wealthiest company became arguably the single most influential book on business ever published. By the time Wanda broke in as a cub reporter, however, the turn-of-the-century muckrakers were history. Since Wanda had not studied much journalism, it is unclear whether Tarbell, exception that she was, influenced her. Wanda had no obvious role models.[32]

That said, the craft of political—as opposed to business—commentary was well developed. The 1940s were a golden age for Washington columnists, with Walter Lippmann and Drew Pearson in their prime. Syndicated columnists were free to rove and sometimes cross the unmapped boundaries between reporting and editorializing, with license to frame their reporting in an interpretive context. Such commentary became a highly influential journalistic form, eagerly consumed by the political cognoscenti. As for business commentary, the *Wall Street Journal* published columns on Washington policies and international developments, but not on industry issues. Business magazines had in-depth articles, profiles, and features, but no regular commentaries. At the *Journal of Commerce*, Emma Doran and several male reporters wrote occasional columns, but not with the frequency, depth, or influence that Wanda's commentaries had during her decade at the paper.[33]

In effect, Wanda had to cut her own path in two ways—as a female business reporter and as an industry columnist. But contrary to Kiplinger's dismissive comment to Eileen Shanahan in the mid-1950s, Wanda excelled at reporting "inside information"—and remained a respectable woman. As a result, she herself, without setting out to do so, would become a trailblazer that other business reporters—both male and female—would follow.

INDUSTRY IN TRANSITION

At midcentury, the petroleum industry underwent considerable upheaval. In 1940, the United States dominated the international oil scene through the sheer number of barrels it produced—70 percent of total world output, excluding the Soviet Union and its satellite states. The American price thus determined the world price. But rapid industrial expansion during World War II and then a surge in gasoline demand turned the United States from net exporter into net importer in those postwar years. Oil quickly overtook coal as the country's primary source of energy. Short-term fuel shortages sparked rapid price increases through 1948, as oil prices rose to double the level of three years earlier. This disorderly transition from wartime rationing to peacetime boom led to problems for both industry and government.

Different segments within the industry vied with one another on almost every issue, from import controls to production ceilings. Among the vertically integrated companies—those that explored, produced, refined, marketed, and distributed domestic and foreign oil—the five dominant ones were Texaco; Gulf Oil; and three successor companies to Rockefeller's Standard Oil—Standard Oil of New Jersey (known as Jersey, later renamed Exxon and then Exxon Mobil), Standard Oil of California (known as Socal, later renamed Chevron), and Standard Oil of New York (known as Socony Mobil). These five U.S. multinationals, along with Anglo-Iranian (later British Petroleum) and Royal Dutch/Shell, were the leading international oil companies known as "the majors" and later dubbed "Big Oil" and the "Seven Sisters." Other oil companies—generally known as "the independents"—had narrower interests. These consisted of about twenty purely domestic, vertically integrated companies; two hundred refiners, who turned crude into gasoline, jet fuel, and other products; and thousands of politically powerful smaller independent producers. The distribution segment also accounted for hundreds of pipeline companies, tanker fleets, and whole-

sale distributors, plus thousands of independent service station owners.[34]

Publicly, these companies all claimed allegiance to the competitive,
free-enterprise doctrine, but in practice, each group benefited in some
way from government regulation. State and federal authorities provided
for oil output controls ("prorationing"), leased public land on favorable
terms, and granted generous tax breaks—policies developed because of
oil's boom-and-bust cycles. When huge discoveries in the 1920s led to a
price collapse, oilmen temporarily overcame their distrust of government involvement and pleaded for a bailout with measures to help
rationalize the market. New Deal planners obliged: Washington, in cooperation with the states, assigned each producing state a "voluntary"
quota based on expected demand, and new tariffs protected domestic
producers from a potential influx of foreign oil.

In effect, the government did for producers what they could not do
for themselves: limit output in order to keep prices up. State regulators,
sensitive to industry concerns, allowed producers to exert considerable
influence. Texas producers usually managed to resolve their differences
through their regulatory agency, the Texas Railroad Commission (originally set up to rein in the railroad barons). Informal business deals fortified those public arrangements, so both the regulatory structure and
these informal deals promoted a kind of oligopoly that stabilized the
market. Still, enough competition remained to bring about a gradual
decline in real prices after 1948. Once the United States became an importer, economist M. A. Adelman notes, a "huge gap grew between the
potential (monopoly) price" and the actual level. A truly successful
oligopoly could have dictated a higher price.[35]

Washington's postwar relationship with the petroleum business,
both domestic and foreign, remained contradictory as difficulties
stemmed from fundamental conflicts in ideology and a lack of understanding of the cyclical problems of the oil business. A collectivist strain
in America viewed big business as inherently bad. Practical New Dealers accepted large-scale industrial capitalism, but sought government
regulation to control its excesses and assure an equitable distribution of
its benefits. Free-market purists, on the other hand, believed that price
optimized the most efficient flow of goods.[36]

Self-interest, however, distorted and subverted the industry's free-market claims. Though tireless advocates of laissez-faire capitalism, the Seven Sisters much preferred market stability to genuine competition. The independents, worried about the majors' foreign oil, sought government refuge through import quotas and used their political clout to get congressional hearings on charges of conspiracy among the majors. Meanwhile, Justice Department antitrust lawyers, opposed to power concentration in big business and unsympathetic to the complexities of the industry, were quick to suspect collusion. None of these groups readily acknowledged the fact that federal and state regulations helped protect the industry—and the public—from the twin demons of the oil business: scarcity and surplus. These differences contributed to wide-spread public distrust of the U.S. oil industry.[37]

SCOLDING THE SECRECY

Having grown up in the oil industry, Wanda had internalized a sympathetic view of both the oil business and the importance of market capitalism. Her upbringing had given her a thoroughly Western view of the benefits of modernization—especially oil-powered modernization—and her international background gave her a broad perspective on the significance of oil to the world economy. Price controls may have been essential during the war, she believed, but not in the late 1940s. She disagreed with those who argued that the petroleum industry was a monopoly. The domestic fuel oil market she covered was highly competitive, whereas on the international scene, the majors needed to band together, she believed, to provide enough capital to develop oil fields and pipelines in logistically and politically risky conditions.[38]

The legendary secrecy of the major oil companies, however, made it a tough industry to cover. In the tradition of John D. Rockefeller, many executives maintained an aura of mystery about how they ran their affairs and influenced prices. Any effort to expose or explain the inner working of the business, particularly to reporters, constituted a threat, a violation of proprietary information. Only a few among the new generation of leaders, such as Leonard McCollum, president of Continental Oil (Conoco), sought to improve the industry's public image. As for

Standard Oil of New Jersey, the primary inheritor of Rockefeller's legacy, its long-standing policy of silence toward the press was, according to its board, "designed to prevent competitors from obtaining trade secrets." After fielding charges of Nazi collaboration during World War II, Jersey set up a public affairs office, but it had little effect. Many Americans remained wary of the business as a whole.[39]

Working the oil beat for the *Journal of Commerce* made Wanda keenly aware of the industry's insensitivity to its public image. In a 1948 column, she criticized as "wholly inadequate" the industry's efforts to explain its 1947 jump in profits. Her sympathies were clear, but so was her scolding. "Bigness," she wrote in another commentary, "is not all white, but neither is it all black." The majors were "far from perfect," but their financial strength gave consumers inexpensive oil and stronger national security, while their economic concentration had not prevented smaller companies from growing quickly since the war. The public was unaware of these realities, so Wanda issued a bold warning to the industry: oil companies could stay silent only at their peril.[40]

Two columns in 1948 caused a particular stir, especially in boardrooms. Both involved Wanda taking what she called an "unofficial little poll." The first, "A Neglected Front Yard," started with Wanda's summer vacation, a "motoring trip" from New York to Miami. "We were just tooting along the highway when she got the idea," Jaqua recalled. Because oil-company credit cards were a novelty, many cardholders were still not sure how to use them. Could service-station attendants explain how they worked?[41]

Wanda recommended her experiment to anyone concerned with the public image of the oil industry. "Just ask a few questions when you fill 'er up," she wrote. "It's an eye opener.... If you want to hear—from the horse's mouth—that the oil industry is just one vast monopoly," nothing but subdivisions of Rockefeller's old Standard Oil, "then talk with your local service station dealers." When she asked managers at six different Esso (Jersey) stations whether she could use another oil company's credit card, she was told, "Sure." How so? "Oh, they're all the same company, same thing as Esso," came the answer—every time. Even her card from Phillips, the independent oil company? "Sure, same thing." Time and again, she got the same answer. One gregarious manager even fetched a credit-card chart to prove how big "his" company was. If it's all one company, she inquired, why does it use so many names? "For tax purposes," came the prompt reply. In twenty-eight stops, she did not meet one manager who could explain the industry's credit-card exchange program. Managers did not know which companies were affiliates and which were separate companies whose credit cards were honored only for customer convenience. Is it any surprise, Wanda concluded, that "the general public thinks the oil industry is a monopoly," as opinion surveys had shown?[42]

A number of papers and journals reprinted the column, but Wanda was her own best publicist: she sent the article to many oil company presidents. Gulf Oil's chief acknowledged the problem and sent the column to his marketing department. Indiana's vice president told her he found it "very amusing," but Monroe ("Jack") Rathbone, president of Esso Standard, Jersey's refining and marketing affiliate, and later Jersey's president and chairman, did not. The column was "a bit personal" for him, since Esso had started the credit card exchange program. To oil company public relations staff, she later learned, the article proved "earth-shaking."[43]

She then undertook another unscientific survey. Wherever she drove that fall to Atlantic City, Buffalo, Chicago, or her husband's hometown in Indiana, she asked service station managers what they thought about their industry. If senior oil executives tried the same thing, like the "very wise Caliph who disguised himself as a beggar and mingled among his subjects to find out what they really thought of him," they would be shocked. Most dealers, those in direct contact with America's motorists, resented "being treated like a 'greasemonkey'" by their company, which seemed only interested in sales volume. Questioning dealers at random, Wanda found that a few gave seriously incorrect answers and that more than two-thirds were at best "misinformed" about the oil business. The oil industry, she concluded, could not afford to neglect, much less demean, its front-line workers.

The two columns got results. Several companies, including Jersey, reported taking action to improve communication with service station managers.[44]

After this unexpected success, Wanda challenged the politically powerful trade group of the independent oil companies for claiming that crude-oil prices should rise because of an eightfold jump in prewar exploration costs. Using a range of statistics, Wanda argued that they had vastly overstated their case. Yes, prewar finding costs had risen, but at a much slower pace than the group said, which exposed the entire industry to charges of false claims. In a typical response she received, Gulf Oil's cautious president, Sidney Swensrud, praised Wanda in a "confidential" letter, saying that he, too, had complained about these misleading statistics.[45]

Through these columns and anecdotal surveys, Wanda found ways of posing questions and critiquing the industry without alienating its leaders—at least most of the time. These postwar oil executives, though more open than Rockefeller, were notoriously prickly about criticism. "Oil officials seem overly sensitive to any indication that the industry isn't perfect," the State Department's oil expert, Richard Funkhouser, told his colleagues in 1953. "Emotion, pride, loyalty, suspicion make it difficult to penetrate to reason."[46] As an industry watchdog, Wanda had managed to chip away at its long-standing façade of secrecy.

MIDDLE EASTERN INTRIGUE

Of all the topics she addressed in her columns, Wanda's favorite was the Middle East, but her father's job complicated matters for her at times. Eugene was Mobil's leading geologist by the late 1940s, and he was deeply engaged in his company's efforts to acquire a stake in the Arabian American Oil Co., known as Aramco. This Texaco-Socal joint venture owned Saudi Arabia's sole oil "concession"—a contract giving the companies exclusive exploration and production rights to the entire country. The deal had, by then, become the envy of the other majors: oil potential in Saudi Arabia was proving to be gargantuan. Eugene must have talked about it with his daughter, but, anxious not to jeopardize his position, Wanda claimed she never used him as a source.[47]

Once, however, Wanda did cite Eugene in a story. In December 1946, news broke that Texaco and Socal had invited Jersey and Mobil to buy shares in Aramco. In a page 1 story, she reported the deal in greater detail than her competitors did and revealed exclusive information on why Texaco and Socal were bringing two of their biggest competitors into "the largest and most fabulous oil pool in the world." Texaco and Socal would likely sell a 40 percent stake, a key motive for which, she learned from "certain quarters," was partly political. The two companies needed additional investors to spread the risk and gain strength through numbers. The State Department would back the deal because it would "ensure adequate financing and development." Wanda proved correct on all counts. Curiously, given the sensitivity of the issue, she cited the names of the Mobil and Jersey executives who "left here last night by TWA plane for Dhahran, Arabia, to look over the concession"— including "E. Jablonski."[48]

Wanda later admitted getting indirect confirmation of Eugene's involvement from her mother. When she called their home in New Jersey and learned that Eugene was leaving that evening for the Middle East, Wanda called the airport and bluffed her way into confirming her father's flight schedule. With that tip as confirmation, Wanda pried information from other sources, a skill for which she became notorious.[49]

The Aramco negotiations gave Wanda considerable fodder. By watching import activity in May 1947, she landed another controversial front-page scoop: Jersey and Mobil, she learned, both wanted to import Persian Gulf crude for the first time. Two days later, however, Jersey reversed its decision. After "widespread publicity" about her story, which she reported she had "cleared" with company officials, Jersey shelved the plan without explanation. By saying her scoop had been cleared, Wanda underscored the significance of Jersey's turnaround. Seeking clearance from sources in the postwar years was not unusual, but in reporting the change, Wanda sidestepped the central issue: why Jersey reversed its decision. By avoiding this obvious question, she indirectly acknowledged her need to maintain access to sources, particularly at Jersey, the industry's behemoth.[50]

Protecting sources was a key component of Wanda's modus operandi. She did not believe journalists should report all news at any cost; she was in her career for the long haul. By this time, her goal was to be the country's best reporter on international oil. She knew she antagonized some sources with her investigative reporting, but she was also willing to withhold information temporarily or write elliptically to ensure continued access. "Wanda would do anything to get a good story for the paper and get it right," Jaqua recalled. "She was always full-speed ahead, always in forward motion. But it was really important for her to hide the sources of her stories. She was great at disguising them." Sometimes, Jaqua said, she would imply that her source for a scoop on a certain company came from that company, when she had actually gotten the tip from an unnamed competitor and only confirmed the information with the original company. She struggled to balance the need to protect her sources with the need to publish breaking news. "Wanda was not someone who wrote things down," Schwartz recalled. "She stored so much in her head. She internalized so much. She felt she had to hold back a lot because she got so much inside, off-the-record stuff."[51]

VENEZUELAN VENTURE

Following her early success covering the majors' growing interest in the Middle East, Wanda became even better known on the international oil scene after a 1948 trip to Venezuela, where she wrote seminal articles about the strategies of Juan Pablo Perez Alfonso, the most original thinker on oil issues in the developing world, and later the intellectual founder of OPEC. These pieces gave the majors reason to worry.

Venezuela was then America's most important source of imported crude oil. Prolific supplies from Mexico had dried up when it nationalized its industry in 1938, after years of frustration with U.S. and British companies. When Caracas pressed for a substantial revision of contracts in 1943, the majors, led by Jersey, Gulf Oil, and Shell, were more conciliatory than they had been five years earlier in Mexico. Jersey in particular had much at stake. Its Venezuelan subsidiary, Creole, was that company's best source of foreign oil and profits. In 1943 these majors agreed to the principle of equal partnership—a fifty-fifty split in profits between the companies and the government—but after accounting and other adjustments, Venezuela got the short end of the stick. The real split was about sixty-forty.[52]

When disaffected young army officers and politicians overthrew a weak military regime in 1945, the new president, Romulo Betancourt, not only vowed to lead Venezuela toward democracy but also pledged to transform the oil industry. He brought in Perez Alfonso, an economist, law professor, and congressman, as minister of development to draft a new policy. By asserting some control over the foreign companies, Betancourt, Perez Alfonso, and their Acción Democrática party provided the basis, as historian Daniel Yergin puts it, "for redefining the relationship between oil companies and producing countries, between tenant and landlord around the world—as well as the methodology for reallocating rents."[53]

In the late 1940s, these Venezuelan reformers were little known in the United States. Press coverage of them was limited to brief wire-service reports. For U.S. oil companies, however, the Betancourt government did not inspire confidence: it immediately imposed a surtax on the most profitable oil companies. Despite their "grave concerns," the majors did not strongly object. Venezuela was still a cash cow for them. Thanks to the revised concessions of 1943, Jersey, Gulf, and Shell had expanded production dramatically, and their profits had soared. With the 1945 coup, the companies feared that the new government would nationalize the industry outright, so the simple surtax was something of a relief.[54]

Wanda wrote enough stories about Venezuelan oil in 1946 and 1947 to attract the attention of the scholarly Perez Alfonso. He accepted her invitation to write for the *Journal of Commerce*'s special 1947 oil issue, although the result was a rather bland statement of his goal: boost oil revenue to improve Venezuela's meager standard of living. Eager for more attention in the American press, Perez Alfonso invited Wanda to Caracas to attend the inauguration in February 1948 of Betancourt's successor, Venezuela's first democratically elected president, the novelist Romulo Gallegos, and to hear his plans for the Venezuelan oil industry. To extend the invitation was a somewhat risky move. Wanda had expressed considerable interest in Venezuela, but her commentary had favored the majors, not Perez Alfonso.[55]

Wanda's trip to Caracas proved decisive. During her two-week visit, she delved into the Venezuelan oil scene at a frenetic pace, packing in dozens of interviews with American and British oilmen and diplomats and Venezuelan officials. But she did not hurry Perez Alfonso, who met her at the airport and surprised her with an invitation to some inaugural festivities. Flustered, she admitted that she had brought only business clothes, so Perez Alfonso, although ascetic by nature, arranged for a shopping tour and paid for her formal dress. John Pearson, a *Business Week* editor who later heard this story from Wanda, thought it an improper gift, but "back then it was no big deal"; standards for journalists "accepting favors were much looser then." Small gifts or free dinners were not going to change Wanda's attitude. She was talking with all sides, though she gave Perez Alfonso the most time.[56]

The resulting stories drew considerable attention in both Venezuela and the United States. Her articles were reprinted in several publications and quoted in more. They prompted letters from company chairmen and were read "with great interest" by government officials, editors, and diplomats in Caracas and Washington.[57] Venezuelan oil was "vital" to Americans, she wrote in her first column, and the newly elected government was genuinely interested in private foreign investment. Contrary to some U.S. press reports, it was not "in bed" with radicals. The New York journal *The Nation* approvingly noted this assessment from the "petroleum specialist of the *Journal of Commerce*" and quoted Wanda at length.[58]

What set off alarm bells at the major oil companies, however, was her page 1 feature on Perez Alfonso in March 1948. This article, said Alirio Parra, Perez Alfonso's assistant in the late 1950s and later the country's oil minister, "shows Wanda's genius." In it she gave "a great synthesis" of Perez Alfonso's views. His boss, Parra said, was long-winded and professorial, "not that coherent. He didn't spell things out like that. Wanda had to piece this together herself. He was the type to ramble on and on and on. She would have had to think a lot about it to put it together." Rereading this story nearly fifty years later, Parra said it was remarkable that "there's nothing favorable toward the majors in this article." Indeed, "Wanda was responsible, without a doubt," Parra said, "for giving Perez Alfonso his first international recognition."[59]

As "prime architect" of the government's oil policy and "one of the most able and profoundly idealistic oil ministers in Latin American history," Wanda wrote, Perez Alfonso was determined to find a more equitable division of profit, while still consulting with foreign oil companies so as not to kill the "goose that lays the golden egg." But Wanda then gave specific information about just how hard Perez Alfonso was willing to squeeze the goose: he would go after a genuine fifty-fifty profit split and stop giving out more concession grants, since current acreage under contract was not sufficiently explored. Perez Alfonso readily admitted his policies favored the majors over the independents, because they "have more to offer," with better wages, revenues, and profits.

Then came the coup de grace. After soothing comments about partnering with foreign companies, Perez Alfonso outlined a completely different conception of the oil business. He wanted Venezuelans to get directly involved as investors and managers, not just as oilfield workers. In addition to receiving royalties and taxes from foreign oil companies, the Venezuelan government wanted for the first time some payment in oil itself, which Venezuela would, through local companies, sell on the international market. His country had learned from the Mexican example "that private companies can do the job better," Perez Alfonso said. Even though he rejected outright expropriation, he hinted at a new, more profitable way for his oil-rich, technology-poor country to gain control of its oil business: a step-by-step takeover through a mixed public and private effort. Perez Alfonso was beginning to articulate a gradualist approach that later became the basis for oil-producing governments' efforts to take control of the industry. More revenue was not the primary issue. He wanted Venezuela to be able to manage its own oil business. This aim, though not completely spelled out, was at the heart of Wanda's assessment: with the prodding of an idealistic but pragmatic professor, Venezuela's oil industry was "undergoing a quiet but profound revolution."[60]

Wanda quickly learned about the article's impact among those who understood its implications. Perez Alfonso wrote to say the article "reflected very well the overall situation and outlook now existing in Venezuela." He expected to come to New York soon: "Perhaps then I will have a chance, too, to see the Czech national dances you mentioned and try those special dishes you wrote me about."[61]

The oilmen were less sanguine; they clearly got the message. It was "quite a shock" for them to find their suspicions "confirmed in cold print," wrote Robert Siegel, a Socony Mobil executive in Caracas. "The only out appears to be Perez Alfonso's statement that these views represent his present thought on the subject and that they may change if they are proved impractical." Siegel added that he had just visited the U.S. ambassador, "and found that this particular clipping was being presented to him for his reading."[62]

SECOND ONLY TO A QUEEN

Shortly after the article reached the desk of the chairman of Standard Oil of New Jersey, Wanda received a query: Would she be interested in an interview with Jersey's board of directors at company headquarters in New York? Wanda was stunned. Getting an interview with the entire board of the world's largest multinational was almost unheard-of, given Jersey's aversion to the press, and she was a twenty-eight-year-old woman.[63]

When Wanda arrived at the Rockefeller Center headquarters, she approached the austere boardroom with trepidation. "Her knees were knocking," *Business Week* editor John Pearson remembered her saying. "She was so intimidated." The chairman, Frank Abrams, however, was a cordial, easygoing man. "Unlike so many of his peers, Abrams showed no fear of the press," company historian Bennett Wall notes, whereas Jersey's president, Eugene Holman, was "painfully diffident." But he, too, was there. He had a particular interest in Venezuela since he had been president of Jersey's subsidiary Creole. Holman had also met Wanda before; she had interviewed him, printed one of his position statements, and prompted Jersey's board to reverse a decision on Middle East imports in 1947.[64]

Wanda quickly realized that she was not the only one who would pose questions. As a friend from Creole, Ted Cook, later wrote to her, "The grapevine reports that recently the Jersey Board of Directors turned the tables and interviewed you." They quizzed her about what she had learned in Caracas, he said, because of her "accurate and comprehensive picture of the situation." She certainly made an impression. After the interview, Abrams wrote to "express our appreciation to you for giving us first-hand impressions of a disinterested observer" of the Venezuelan scene.[65]

Although Wanda never wrote about her interview with the Jersey board, word got out. From then on, she had access to the chairmen of all the major oil companies. If Jersey's chairman was talking with her, other chief executives wanted to meet with her, too. Wanda used this leverage for the rest of her career. Because these companies were expanding so rapidly in the postwar period, particularly abroad, they were becoming more sensitive to their public image and their need for intelligence-gathering on an international level. An investigative reporter such as Wanda could be a valuable independent source of information. Their need created her opportunity.

Wanda took advantage of the situation, but she was not merely the right person at the right time. Alternately blunt and tactful, she had a way of identifying and synthesizing the most significant issues. Poised and seemingly self-assured, she did not fit the stereotype of the fast-talking, slouchy-suited business reporter. She knew she had invaluable access that was denied to most business reporters—she had gained and kept the trust of some extremely powerful men, and she did not flaunt it. As long as she did not compromise her position as a journalist, she was willing to talk about her insights.

Indeed, a Jersey public relations official told the radio host of "The Betty Crocker Magazine of the Air" that Wanda should be invited to speak on the radio show "because she is young, can talk ad lib, and has a number of colorful anecdotes to tell." Incidentally, he added, "she is a housewife in the bargain." But in the end, the session at Rockefeller Center was not only interesting; it was historic. Wanda Jablonski was the second woman ever to address the board of directors of the most powerful company in the world. The first was Queen Marie of Romania.[66]

WASHINGTON GADFLY

Along the way, Wanda discovered that she rather liked controversy and attention. The greater her success, the more she overcame her shyness, and the larger her targets became: after her high-profile trip to Venezuela, she delved into Washington politics. During a series of hearings in 1948 into the alleged monopolistic practices of large oil companies, Wanda denounced Nebraska's Republican senator, Kenneth Wherry, chairman of the Small Business Committee, for charging that the industry was engaged in a "vast conspiracy" to restrain production and boost prices. The senator supported his allegations with a special report from his oil counsel, Paul Hadlick. "Where Mr. Hadlick got his amazing 'facts' is a mystery to everyone," Wanda countered, citing her own facts to refute each charge. Hadlick "has long been known in the industry as a specialist in petroleum marketing," not production, and was therefore no more of an expert on domestic oil production "than a tank truck peddler in Maine."[67]

Her sarcasm did not go unnoticed. Wanda got a letter from the committee bearing a threat that was veiled but, to Wanda, quite clear: "Will you advise by return mail if any of your immediate relatives or parents are employed by any of the oil companies and if so, state their names and also company connections." Wanda was not about to be intimidated. It was obvious, she wrote in reply, that the committee already knew about her "parent" geologist. "Are you questioning the integrity of my column? Or inferring that the views I publish are not my own?" Her paper's management, she wrote, was awaiting clarification. Her letter files contain no response.[68]

Although dismissed as a gadfly by some politicians, Wanda gained influence in Washington. She heard that some of her columns were under scrutiny at the White House, and it was "well known," an Interior Department official told her, "that your editorial approval is not easily won." She even provoked ten days of controversial hearings in July 1953 with an article on whether collusion among the oil companies had led to a retail price increase that spring. Wanda herself was named as instigator of the price hike, and this time she responded in her columns by scolding politicians as well as oil executives for misleading the public. Her impact was significant. She was not only a subject at the hearings because of her ostensible role in the price increase, but through her columns, she also contributed to defining the issues, highlighting inconsistencies in testimony, and clarifying the controversy for readers. That she was a woman prompted interest but also derision.[69]

The hearings started out with a remarkable claim. New Jersey Republican Charles Wolverton, chairman of the House Commerce Committee, declared in his opening remarks that one of Wanda's columns had provoked the price increase, and he wanted to know why. To him, higher prices were unwarranted and inflationary. The problem, he said, could be traced back to Standard Oil of New Jersey. At a Wyoming conference in May, Jersey's Rathbone had said that his company, which had resisted price increases for more than five years, would no longer oppose a hike, but the press did not notice this surprising statement until Wanda's column appeared in early June. "The crude price rise," Wolverton charged, "followed shortly thereafter." Was Jersey's policy change, as explained by Wanda, an open invitation from the country's largest oil company for someone else to start raising prices? Wolverton wanted answers.[70]

How Wanda learned about this policy shift is unclear. She did not attend the conference, so someone may have tipped her off, but it was the way she reported and interpreted the speech that mattered. In a front-page column, she explained that since Rathbone's statement was in the text, not an off-the-cuff remark, it actually "reflected the thinking of Jersey's management." When Jersey executives gave speeches, she noted, "every word" was "about as carefully written and screened for policy as the original charter of the United Nations." Jersey had resisted price increases for five years because it hoped stable prices would keep Washington from imposing price controls. Rathbone's statement, however, was Jersey's signal that a market-driven price increase was acceptable. This notable shift in its position removed a stumbling block for smaller producers who wanted higher prices given the higher demand. That change, she warned readers, made Rathbone's statement front-page news.

But she had a caveat: Contrary to popular belief, what Jersey wanted, Jersey did not always get. In 1947, it had objected to a 50-cent-per-barrel increase, and yet, because of extreme shortages, it was eventually forced to match its competitors at the higher level or lose suppliers. Then again, its opposition in 1948 to another round of price increases did succeed, as smaller companies eventually rescinded their initial price hikes. In early 1953, oil supplies were fairly ample, but if the market tightened up by autumn, as expected, a price increase would be more likely, Wanda concluded, thanks to Jersey's shift in policy. Six days later, the price hikes began—and stuck.[71] Industry critics charged collusion.

The committee probably asked Wanda to testify, friends of hers later recalled. But if it did, she declined, as she did all congressional requests for expert testimony throughout her career. Aside from a fear of public speaking, she saw no need to jeopardize her job. Prompted by a set of questions that Chairman Wolverton wanted witnesses to answer, Wanda replied on her own terms — in print. Since state agencies controlled production ceilings, she explained in a July 7 column, the oil market was partly regulated and partly unregulated. As a result, contradictions in the law of supply and demand did crop up, and refiners were sometimes caught in the middle. In this instance, somewhat tighter crude oil supplies made a price increase understandable. Gasoline supplies were ample but not overstocked, so the higher prices were likely to stick. Heating-oil supplies, however, were excessive, and prices were already declining. The mixed regulatory situation was causing these contradictory price trends.[72]

Wanda's explanations made sense, but industry executives who testified the next day did not. Hines Baker, president of Jersey's refinery affiliate, presented "nothing but generalizations," Wolverton complained. Baker denied that Jersey had tried to influence prices, but his long-winded explanations only confused committee members. One frustrated congressman accused the executives of "living in an ivory tower." A senior Conoco official who watched the proceedings with disgust told his company's board that the "bewildering and unconvincing" executives should just have quoted Wanda, as she did a "much better job" than did Baker of explaining the "intricacies of the oil markets." He praised her July 7 piece as the "best and clearest presentation of a complex subject that I have ever seen."[73]

Wanda followed with two more scathing commentaries. On July 14, she asked a committee member, John Heselton of Massachusetts, why he blamed the industry for a rise in gasoline prices in his home state — the first increase in more than three years — when two-thirds of it actually came from state and federal tax increases. Sixteen other states had also raised gasoline taxes. Congressmen Wolverton and Heselton, she wrote, "showed a patently hostile attitude" toward industry witnesses at the hearings, "playing for the headlines with the old political game of baiting the 'bad big-company wolf.'"[74] But she saw in the hearings more than just a "biased forum." The oil industry was also very much to blame. Executives clouded the issues by claiming the market was fully free, when they should have frankly admitted that the market was partially controlled by state commissions, that some regulation was vital to the national interest, and that economic incentives were needed to ensure adequate supplies. With that kind of candor, she wrote, "the oil industry's position would make a lot more sense."[75]

This column, however, got her into further trouble. Heselton used it to try to corner a fearsome witness, General Ernest O. Thompson, chairman of the Texas Railroad Commission. Would Thompson admit, Heselton asked, that the oil market was regulated through production ceilings from state commissions? No, that was false, Thompson replied. Heselton then proceeded to read part of Wanda's "Petroleum Comments" from that morning's issue, July 23, omitting her charge that he and Wolverton held a "patently hostile attitude" toward the industry. (When he got to that point, Heselton turned to the chairman to say, "And then she pays respects to yourself and myself.") Heselton liked her advice for the industry: its leaders should admit that producer states regulated the market by adjusting production levels to industry projections of market demand. How could oil states, Heselton asked Thompson, permit this kind of market-fixing?

Thompson's first line of defense proved, once again, how difficult it was for a female business reporter to be taken seriously: "It is a lady saying it; it is not correct." Heselton protested that Wanda was nonetheless "recognized as a pretty competent person," and Thompson played along: "She is a very fine writer." She was not an oil expert, he implied—just a writer. Yet "she is not unfriendly to the oil companies," Heselton countered. Frustrated with the subject, Chairman Wolverton said, "We certainly have recognized the lady considerably during these hearings." To which Thompson again pointedly noted, "She is a very fine lady."

Then Thompson launched into a full-blown attack. Wanda Jablonski, he said, was wrong to say that Texas was, in effect, fixing the market. "We don't do that in Texas," the bureaucrat added. "We just enforce the law, nothing more." Texas regulated production only "to prevent waste, and we do regulate it to market demand, but that is our law." As for

Wanda herself, "I think she is a lovely person," but he disagreed with her. After more protestations, he concluded, "I am just an administrator." Thompson, in effect, admitted that the Texas commission set regular ceilings on the state's oil output to ensure that a limited supply could meet market demand, but he refused to acknowledge that these decisions affected prices. Instead, Thompson twisted Wanda's lucid explanation of a partly regulated market into an accusation of "price fixing."[76]

The lady in question decided not to respond, at least not publicly. She had already made her point several times in print. Most of her important readers knew by 1953 that she was not a "Bill," and she knew they paid serious attention to what she wrote. This hard-won sense of assurance that her voice was now heard, this taste of power and influence in the face of skepticism and derision, would lead her to venture further afield. That summer she managed to persuade Luedicke, her editor and mentor, to send her to the most controversial place in the international oil world: Iran. Despite her fear of flying, she knew she had to see for herself what was going on there.[77] The future of world oil hinged on what would happen in the Middle East.

3.

Iranian Intrigue

"Chances are the Americans will be the goats this time, just as the British were the last time," wrote W. M. Jablonski in the *Journal of Commerce*, dateline Tehran, February 8, 1954. This was Wanda's conclusion after spending two months in Iran reporting on its oil crisis. Her sardonic tone marked a real change from prior commentaries written in New York, which were generally supportive of the British in their showdown with Iran. Senior Western diplomats and oilmen did not expect this sort of commentary from someone they had urged to go to Tehran to "educate" those "unworldly" nationalists about the "hard facts" of the oil business. And for those Iranian nationalists trying to salvage some dignity despite their nationalization debacle, it was, one of them said many years later, "exactly word for word" how they felt.[1]

Wanda enjoyed surprising people. It was one of the things she liked best about being a reporter, whether she was getting a scoop or correcting a misconception with new information. On her way to Tehran in December 1953—her first visit to the Middle East since childhood—she stopped in Baghdad for a few days, just long enough to find and dispatch the first surprise from her trip. The oil companies had kept quiet about several huge discoveries in the Persian Gulf region—until Wanda appeared on the scene and made them page 1 news. When she quizzed officials from the Iraq Petroleum Co., a British, American, and French joint venture, they conceded that they had "barely scratched the surface" of the newfound reservoirs. One elated geologist told her that she could safely say that more oil has been found "than the world can possibly use for a long, long time to come." Despite this excitement, Wanda was soon ready to move on to where the real Middle East oil story lay: Tehran. There, shockwaves from Iran's nationalization of its oil industry still reverberated.[2]

Arriving in Tehran just before Christmas, she found it "a city of drab monochromes, of mud-colored bricks and high mud walls, cuddling up against the glistening white Elburz Mountains." She set to work immediately, hastening from government ministries to foreign embassies, from evening receptions to late-night hotel bars. She queried and probed. Along the way, she lined up exclusive interviews with cabinet ministers, traded information with ambassadors, and talked to high-powered men—Iranian and Western alike—who were more than a little surprised to find themselves discussing oil politics with a young, rather attractive woman. "I thought that any woman journalist would be square and have a mustache," remembered one Iranian. She listened intently, and the longer she stayed, the more nuanced her perspective on the developing Iran story became.[3]

Although Wanda arrived with Western proclivities, her sensitivity to nationalist aspirations allowed her to see and hear and understand more than other foreign reporters did. The Iranians, she came to realize, needed to better understand how oil markets worked and why the international companies were important in helping them find, develop, and market their reserves abroad. The oil companies, on the other hand, needed to understand why they should share control over production and development with the host country, bringing Iranians into positions of real responsibility. Her coverage, as it turned out, influenced both sides.[4]

NATIONALIZATION

The Iranian oil crisis that drew Wanda to Tehran had begun nearly three years earlier. After decades of frustration with Britain's iron-fisted control over its oil and years of trying to negotiate better royalty payments, Iran's parliament, the Majlis, voted in March 1951 to nationalize the country's oil industry. That industry was run by the Anglo-Iranian Oil Co. (renamed British Petroleum in 1954), which, along with its principal shareholder—the British government—had grown rich and powerful off Iranian oil. Indeed, the original rationale for Anglo-Iranian Oil was to secure oil supplies for the Royal Navy, backbone of the British Empire.

In 1909, the company had acquired exclusive ownership of Iran's only petroleum concession, giving it legal title to any oil discovered and produced, plus total control over its development and export. As one historian put it, Iran's leaders "sold their birthright for a pittance." By 1950, Anglo-Iranian Oil boasted the highest production rate in the Middle East: 40 percent of the region's total output. It had also built the world's largest refinery, at Abadan Island on the Persian Gulf. Rather than prosper with the British company, however, Iran remained a mostly agricultural, impoverished country. Britain's most profitable company paid Iran less than what it paid the British government in taxes. Growing Iranian demands for a more equitable contract were simply brushed off.

Anglo-Iranian Oil's high-handed management only added to decades-old resentment of Britain's imperialistic ways. Embittered by this treatment and eager to respond to a growing popular outcry, the Majlis nationalized the oil industry—a symbolic yet very real overthrow of foreign domination. In May 1951, Mohammad Reza Shah, the young titular head of the country, signed the law revoking the British company's concession and creating the National Iranian Oil Co. (NIOC) to take its place. He also accepted the Majlis' choice for prime minister— the charismatic sixty-eight-year-old Mohammed Mossadegh, who had led the fight for nationalization.[5]

When American attempts to mediate the crisis in mid-1951 failed, Britain imposed a boycott, and Anglo-Iranian Oil threatened to sue any company that bought Iranian crude. The embargo worked. The other majors readily complied because they did not want to undermine the concession system on which they, too, depended, and they could quickly make up the shortfall by boosting production elsewhere. Just as they had done with nationalizations in Mexico and Bolivia in the 1930s, the companies closed ranks. Despite genuine rivalry among them, interlocking arrangements made for a common outlook. A threat to one competitor's concession was a threat to all.

Iran held out as long as it could, and by 1952 the country's abundant oil flow had dried to a trickle. Still, Mossadegh would not compromise. The stalemate continued until American policy began to shift in early 1953, when Dwight D. Eisenhower became president. His predecessor, Harry S Truman, who was sympathetic to nationalist movements, had welcomed Mossadegh to Washington in October 1951 after the prime minister made an eloquent defense, at the United Nations, of Iran's right to control its natural resources. However, Eisenhower's foreign policy advisors—particularly Secretary of State John Foster Dulles and his brother, Allan Dulles, director of the Central Intelligence Agency— saw Iran as a crucial bulwark in the great power struggle of the cold war. For them, Mossadegh's defiant stance was intolerable; an independent Iran could easily fall under Soviet Russia's sway. In August 1953, American and British intelligence operatives secretly engineered a military coup that deposed the democratically elected Mossadegh, brought back the shah, who had fled to Rome, and turned the government over to the CIA's handpicked leader, General Fazlollah Zahedi. The true story of how the CIA initiated and directed the coup remained secret for years. At the time, the Western press reported that military officers had rebelled against pro-Mossadegh generals and street mobs had followed.[6]

When Wanda arrived in Tehran in December 1953, the prevailing question was what Zahedi, the new prime minister, would do to settle the nationalization issue and restore oil exports. A bevy of U.S. and British government petroleum experts and oil-company executives had met in London earlier in December to prepare for negotiations with Iran that would restore some kind of foreign management of its oil industry and compensation to Anglo-Iranian Oil. American diplomats were pushing for an international consortium led by the American majors. Anglo-Iranian Oil would be included, but only as a large minority shareholder.[7]

For Washington, Iran was not Mexico or Bolivia. National security needs and the world oil scene had changed dramatically in fifteen years. Instead of accepting nationalization as a fait accompli, Eisenhower's administration wanted a cooperative relationship with Iran—and, most important, a pro-Western outlook. The central issue for the United States, wrote one American journalist just before Wanda's arrival, was not so much gaining access to more oil in this "nation intoxicated with its newly won sovereignty," but "plugging holes in the front line of the non-Communist world, wherever it is weakest." Getting Western oil companies back into Iran was critical. Containment of communism was at stake.[8]

NATIONALISM, NOT COMMUNISM

Wanda developed a different angle. Yes, Iran needed assistance from the majors, but the country was not "intoxicated." Washington diplomats and the American companies would be making a costly mistake if they did not "take full account of Iranian sensibilities," she wrote in the *Journal of Commerce*. If the Americans insisted on a "purely hard-headed business deal" to restore Iran's oil industry through their consortium plan, they would, indeed, replace the British as the "goats."

Americans, she argued, did not understand a central issue for Iran: nationalization was not socialism, as it might be defined in, say, Britain; nor was it communism. "It is strictly a nationalistic concept," she explained, "a distorted, but nonetheless genuine, expression of the country's desire for independence from centuries of foreign political domination." There were good reasons why Mossadegh was "still considered something of a hero" by many people in his country. Any settlement with Tehran should have "a clear, concrete program of 'Iranizing' the industry" to give Iranians a substantive role in management. If not, discontent would erupt again, and the American consortium would look just as bad to the Iranians as the British company had.[9]

An Iranian engineer, Parviz Mina, then a graduate student and later chief of international affairs for Iran's national oil company, the NIOC, was "simply amazed" by Wanda's column. "If I had been able to write an article at that time expressing my feelings about the oil industry, this is exactly what I would have written," Mina said years later. From December 1953 through early February 1954, Wanda talked repeatedly and at length with the NIOC's senior officials and government leaders, such as the foreign minister. Most Westerners, whether reporters, diplomats, or businessmen, Mina said, did not show that kind of interest in the Iranian point of view, let alone write about it.[10]

Indeed, many educated Iranians of his generation had painful, at times humiliating, experiences with Westerners, particularly the British. When Mina was a trainee one summer at the Abadan refinery, he would stand on sore feet for hours in the stifling, mind-numbing heat to wait for an "Iranians-only" bus. Comfortable buses for the British staff, whether managers or maids, came much more frequently. Though Mina already had one British university degree in engineering and would soon earn a doctorate, the British treated him, he said, as though he were in "a lower position even than a servant." Wanda was different, he recalled: "Wanda never gave us the feeling that she was looking at us as second-class citizens." That was why she was "so skillful in getting us to open up."[11]

PRESS STEREOTYPES

American press coverage of Iran shifted during the crisis. In April 1951, most newspapers reflected the State Department's initial position that Britain should accept nationalization as a fact. "The U.S. won't let Britain use force in Iran," the Wall Street Journal reported, and Britain should stop its "19th century threats." But then the British embassy began intensive lobbying to discredit Mossadegh and the Iranian nationalists. In September, a New York Times editorial asked, "Is it really conceivable that the United States should reward Iran for breaking her oil contract....Who is our greatest ally in the defense of the West, Britain or Iran?"[12]

The Wall Street Journal also began to voice its disapproval. Edward Hughes, a roving Wall Street Journal correspondent who had arrived in Tehran in May 1951, derided Iran's decision to nationalize the oil industry over a mere "squabble." "Violence, fanaticism, prejudice and terrorism [were] rampant," he wrote, and communists might take over. The British, who had gone "a long way toward meeting Iranian desires," said that Washington's conciliatory attitude was a sign of weakness. "We have learned from years of experience out here," a British diplomat told Hughes, "that you must be firm with these people."[13] The Anglo-Iranian

Oil Co., Hughes conceded in one commentary, should have offered Iran a better contract, but he could not blame the British for secretly trying to bring down the Mossadegh regime since nationalization was "irrational, illegal and immoral." What's more, a "hysterical atmosphere" prevailed in Tehran. Since it was unlikely that Mossadegh would agree to "emasculat[e] the nationalization program," Britain's chief task must be to help him "crawl back off the shaky limb on which he has so carelessly flung himself." Negotiations, Hughes concluded, must begin immediately.[14]

At the Abadan refinery, the future looked bleak, Hughes said, because of a "wacky little cold war" between the Iranians and Anglo-Iranian Oil. "Poker-faced" Iranians "grappled" for control of "air-conditioned offices, fancy limousines and motor-boats." Preoccupied with these "highjinks" in their "toy" war, these "reckless" Iranians claimed they could run the complex refinery on their own but showed little interest or expertise, so it was unlikely to function without British know-how. To Hughes, it was "one of the weirdest industrial property transfers in history." He made no mention of what the *Jerusalem Post* reported that same month: that living conditions for Abadan's refinery workers were abysmal, with many living in open tents or cramped, unsanitary quarters. Instead, on a return trip to Tehran in September 1952, Hughes reported that the Iranian premier was a "crafty old Oriental bargainer," unwilling to meet Britain halfway. Instead, he was using "stunts" to keep the British at bay and "hocus-pocus" to pump more paper money into his economy.[15]

Other publications were also hostile toward Mossadegh and his followers. "The symbol of fanatic, know-nothing nationalism," Mossadegh had "whipped" Iran into "frenzied hatred" of the British, *Business Week* said, whereas the shah was modern and idealistic. Even though oil was "the great internationalizer," *Fortune* noted, it was not surprising "that irresponsible nationalists should regard the oilman as their particular enemy." The only American oil writer who made it to Abadan in the early 1950s, the *Oil & Gas Journal*'s Dale Duff, focused mostly on technical developments, and though he judged the British evacuation to be "pretty tragic," he wrote that some Iranian engineers had "an amazing ability to keep mechanical equipment functioning."[16]

INSIGHT AND INFORMATION

Wanda was not immune to Western stereotypes of Islamic cultures. Be-fore her trip to Tehran in late 1953, her commentaries on the crisis re-flected the views of the American experts she talked with—senior oilmen, consultants, and diplomats who were trying to broker a com-promise between intransigent personalities in Britain and Iran. She de-picted Iran's leaders as "extremist nationalists" who had a "strange concept of business ethics" and expressed "an almost childlike faith" that the United States would rescue them from their economic straits. Mossadegh had encouraged "the masses" into a "frenzy of fanatic na-tionalism," she wrote, leaving little room for negotiation. But Anglo-Iranian Oil's chairman, William Fraser, drew harsher criticism. Self-righteous and out of touch, he was to blame, she insisted, for most of Anglo-Iranian Oil's problems. The company got into trouble because "it did not integrate itself properly into the country" the way other ma-jors had in Venezuela and Saudi Arabia. Despite the company's mis-takes, however, Wanda argued that Washington could not afford to undermine the oil-concession system so vital to American national se-curity. She did not come to her new "goats" assessment of the U.S. posi-tion until she had spent more than a month in Tehran.[17]

In her first major interview in December 1953, Wanda reported that the new government's foreign minister, Abdollah Entezam, had a "real-istic grasp" of the oil issues and recognized the need for foreign man-agers in Iran's oil industry. In another piece, after itemizing eight problems Iran faced in the international oil market, she wrote that the Zahedi government, by avoiding the "fanaticism, xenophobia and nar-row chauvinism" of the previous one, could develop a modern state thanks to its oil resources. Iran did not have to remain a "static, desper-ately poor agricultural country like Afghanistan."[18] As a Westerner, Wanda took for granted that Iran's goal must be modernization.

But as a reporter, she was out to "discover the ways the Persians are thinking," she told a British diplomat, so she could explain them to her American readers. She also thought she could contribute informally to Western efforts to find a compromise on the nationalization crisis by "putting over a few hard facts" to her Iranian contacts about the intrica-cies of the international oil business. Denis Wright, the respected chargé d'affaires who reopened the British embassy in Tehran in late December 1953, thought the *Journal of Commerce* had dispatched Wanda at the instigation of Herbert Hoover Jr., the State Department's oil consultant and son of the former president. "Through her articles and contacts with Iranian ministers and officials," he wrote in his un-published memoirs, "it was hoped to bring home to them some of the realities of the world oil market where, with increased production in Kuwait and elsewhere, Iranian oil was no longer needed." Hoover, who had met with Wright in London in late December for U.S.–British con-sultations on Iran, "told me to look out for Wanda." Walter Levy, Hoover's advisor during the consortium talks, however, said Hoover had nothing to do with Wanda's assignment; he was only a source.[19]

The American ambassador to Iran, the consummate diplomat Loy Henderson, quickly decided that Wanda could be useful to him in sev-eral ways. Worried about Iranian public opinion, Henderson cabled Washington that much of the "influential" Iranian public remained critical of U.S. efforts to restart negotiations. After reading Wanda's first articles from Tehran, he took several steps to ensure that her views on the "realities" of the oil market reached a wide audience. Because her articles would be "helpful in forthcoming negotiations," he wanted their full text transmitted immediately to him, and he had several pieces broadcast on Voice of America radio in January 1954 and reprinted in Iranian publications, making her name "a household word in the capi-tal," one oil company newsletter reported.[20]

Henderson also arranged for her to give some brief remarks to re-porters from Tehran's leading newspapers over drinks at the home of Henderson's press attaché, Ed Wells. Foreign correspondents were also invited. "Tehran was a very social place in those days," Denis Wright re-membered years later. "There were lots of great parties." Although Wanda did not like to speak in public, she was good at it. For this infor-mal press conference, she sat on the edge of a sofa, surrounded by a sea of men. Wearing a trim, dark suit offset by a white bow, she looked in-tense and nervous, though she also appeared younger than her age (thirty-four), and she was clutching a cigarette, as always.

As reporters leaned forward, straining to hear, she fielded many questions. Some were hostile. Was she an agent of the American

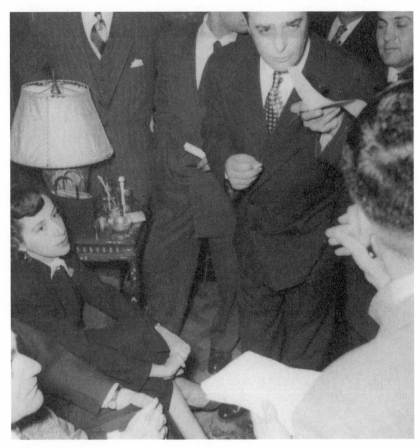

Wanda holding her own informal press conference for Iranian reporters, Tehran, January 18, 1954.

government? "No," she replied. "If my boss reads that, he would fire me." She had come to Iran as a reporter from a commercial New York newspaper to get firsthand information about oil in the Middle East. Asked point-blank whether she supported the nationalization of Iran's oil, she responded tactfully, "We in America know very little about you—that is why I am here to learn."[21]

She explained a number of issues that the Tehran papers found useful. The only companies that could buy large quantities of Iranian oil were the seven international majors that controlled, like it or not, 90 percent of the world's outlets. Iran's "natural" markets were in Europe, not the United States, because Venezuela and Canada took care of America's import needs; and the majors sold a lot of Middle Eastern oil in Europe. To penetrate this growing market, she said, Iran had to establish an efficient, reliable, nonpolitical management of its oil industry.

She was less than candid, though, in responding to a question about whether the "American oil trusts" were preventing the sale of Iranian oil. No, she answered, the majors were not preventing anything. They had more oil than they could use. They did not buy and sell oil on orders from the U.S. government but only made commercially sound business deals. Technically her answer was correct, but she failed to mention that the majors had readily supported Anglo-Iranian Oil's embargo in order to protect their own Middle Eastern concessions.[22]

The Tehran press remained skeptical about Wanda's motives, but they devoted considerable front-page coverage to her comments. The fact that she spoke at an embassy official's home meant that Washington thought her "contacts with newspaper editors will contribute to the settlement of the oil question," noted one paper, *Seday-e-Vatan*. The major oil companies knew that by "sending experts like Miss Jablonski they create public opinion." On the other hand, she "indulged in contradictory remarks" by revealing "that Iranian oil is not after all as unwanted as some try to make out." Several papers approved of her frankness, and a fairly pro-American newspaper, *Farman*, felt confident that she would "make a point of presenting the Iranian public opinion" to the majors. The respected daily *Ettelaat*, which quoted her comments at length, said that because of her "vast knowledge" about oil, great importance was "attached to her views and opinions."[23]

Without mentioning her own possible influence, Wanda noted a change in the Tehran press by late January 1954: local editorials favoring an oil settlement were more frequent. More important to her was that these commentaries marked "the first time in the memory of observers here that Iranian newspapers have ever acknowledged the existence of commercial—as against political—aspects of the oil problem—prices, management and compensation."[24]

Wanda made a significant contribution in "educating the Persians," Wright wrote to her shortly after her departure. They "were not unreasonable, just unrealistic" about the oil scene. An erudite man who later became ambassador, Wright believed that Wanda helped change the views of Tehran's key opinion-makers—both privately and publicly—by February 1954. She deserved credit for having "played a big part in bringing this unworldly people down to earth," he said. None of the other Western reporters in Tehran at the time, he recalled, including the New York Times and London Times correspondents, had "any kind of impact the way Wanda did."[25]

Wanda complemented British and U.S. diplomatic efforts to draw Iran into negotiations with a consortium of foreign oil companies. Her willingness to explain American policy and the oil market to the Tehran press in an official setting did turn her into something of an advocate for the Western position. But the oil companies would have been upset if they had known how much she was helping key Iranians behind the scenes.[26]

PRIVATE GIVE-AND-TAKE

Indeed, the majors might have even called Wanda subversive. During discussions with senior officials and executives of Iran's newly created NIOC, Wanda privately shared information, insights, and even prized documents on the world oil market and on Venezuela's legal reforms. Decades later, a number of these Iranian officials remembered how important she was to them and how much they enjoyed her personal style of interviewing. "Wanda was very sympathetic to our point of view when we felt the whole Western world was against us," said Fathollah Naficy, a distinguished technocrat and director of the NIOC. "She was most helpful to us in telling us especially about people—their past, their present, how well-intentioned they were."[27]

Wanda also helped a member of Iran's Oil Advisory Board, Prince Manucher Farmanfarmaian, the suave descendant of a former Persian ruler and a member of one of Iran's most powerful aristocratic families. The gregarious Farmanfarmaian found Wanda engaging and informative. Distrustful of the British, he credited her with teaching him a great deal about the international oil market. He frequently sought her out at the Park Hotel, where she was staying, and took her on sightseeing trips. "Over dinner...she would say, 'No, the market is *not* like that. It's like this. You're making a mistake.'" She always had "an opinion to pound," but it was sound. A large measure of her success came from her sense of equality, he said. "She was not superior. Not an American. She would write what you say as much as [views expressed by] the other side."[28]

A former prime minister, Hussein Ala, found Wanda equally helpful. Then minister of court to the shah and a central figure in debates over oil policy, Ala not only gained "valuable information" from her but was delighted when she just happened to have a copy of Venezuela's Petroleum Law in her handbag, the latest version of that country's groundbreaking efforts to assert more control over its oil. The British-educated Ala was so taken with Wanda at their afternoon meeting that he brought her home for dinner to meet his wife. In a letter he wrote shortly after her departure, he thanked Wanda for the document, saying her "presence in Tehran was extremely useful." Given how little information Tehran had about other international oil agreements, Ala appreciated her willingness to help.[29]

With these men, Wanda set herself apart from most journalists. Not only was she a woman—and that did cause a stir—but she also had valuable intelligence. In a delicate tango, she passed on insights and information, sometimes secret and always untraceable, with the unspoken understanding that the recipient would respond, at some point, with inside information for her. Having perfected this modus operandi in New York, she engaged Iranian officials in a similar give-and-take with remarkable ease. She "could extract from [sources] their *arrières pensées*" but respect requests for confidentiality, said Farrokh Najmabadi, a World Bank consultant and former NIOC official. She did not just take information from her sources; she gave back.[30]

Because she was such an unusual journalist, these Iranians did not quite know what to make of her. Parviz Mina thought Wanda had an advantage over other Western journalists because she was female: "At least for us Iranians, we treated her with more respect." But despite her disarming "friendliness and wide smile," many of these Iranians simply related to her as though she were a man. "I am confiding to you as man to man," wrote one NIOC contact, Jawad Khalili. Farmanfarmaian also said he could discuss and disagree with her because "she thought like a man." "She was easy to talk with. I had no inferiority complex with her. We could debate, and it was good. She was a friend of mine, like my brother." For him, this was an entirely new way of dealing with a woman, since Iranian women, though more educated than most of their counterparts in the Middle East, were virtually excluded from public life.[31]

One of Wanda's personal features, however, did give her trouble. In his memoirs, Farmanfarmaian wrote that Wanda "had a mind like a steel trap, a nose like a toucan, and a smile that could crack open an oyster. We got along famously." Despite his unfair description of her large nose, this prominent feature and her central European name made some people think she was Jewish, a notion she would hastily correct. Jews not only were distrusted in Iran and in Arabic countries but also had trouble in some oil companies. According to Brandon Grove Jr., Walter Levy, the international oil guru and a family friend of his, "was very hurt over the years because he was Jewish." He had to deal with "prejudice within the companies and mistrust from some Arabs."[32]

Despite her nose and her sex, Wanda successfully cultivated sources on all sides of the off-and-on negotiations between Iran and the consortium of majors, and between the majors and U.S. and British officials. In mid-January 1954, Foreign Minister Abdollah Entezam gave her a second exclusive interview, this time spelling out the Zahedi government's negotiating stance in even greater detail. Any compensation for the nationalized oil properties, he told her, had to be negotiated as part of one overall "package deal," not a separate settlement—a further refinement of Iran's stance. Two weeks earlier, Entezam had told the *New York Times* that Iran would not compensate Anglo-Iranian Oil until marketing was arranged with the consortium. Through these interviews with the New York press, Iran's foreign minister knew he was communicating directly with the New York–based majors without having to rely on U.S. diplomatic missives. For the *New York Times*, however, the story was page 7 news; for the *Journal of Commerce*, both interviews made the top of page 1. Entezam noticed. Six years later, when Entezam wanted to spell out Iran's position on the formation of OPEC, he turned to Wanda Jablonski.[33]

A DIFFERENT VIEW OF ABADAN

Wanda's 1954 reports from Iran attracted attention in New York, Washington, and London, as cable traffic and letters in and out of the State Department and several British ministries attest. Right from the start, diplomats and specialists were struck by the amount of information she was getting from Iranian officials. The American petroleum attaché in London fretted about some of Wanda's scoops, and the oil specialist in Britain's U.S. embassy, John Brook, recorded with "alarm" the estimates in the January 4 *Journal of Commerce* for future Iranian production. "The Jablonski article sounds awfully like some of [Hoover's] propaganda" and, therefore, it might "be difficult to apply the brakes." However, just days earlier, Brook had called an article of hers "constructive and factual." It had raised "a point which no one has yet brought out on this side but which was, of course, on our minds long ago"—namely, the tangled currency issue that would arise from a mixed American-British consortium of companies.[34]

At the majors' headquarters, Wanda's articles from the huge Abadan refinery complex attracted special attention. The oil executives had had little news from there since the last Anglo-Iranian Oil Co. managers retreated in October 1951.[35] Wanda fired off six pieces—two news stories and four commentaries—within a week of arriving at the refinery in late January. Abadan's size and complexity, its isolation and dramatic location on an island near the Persian Gulf, inspired her only feature piece from Iran. "The great Abadan refinery is an awesome and majestic sight, rising out of the desolate desert in a symphony of gigantic towers and gleaming pipes," she wrote. "It stands, in nationalized silence, as mute testimony to the wonders of modern industrial achievement, as impressive in its way as the ruins of not-far-distant Persepolis, that spectacular tribute to a by-gone civilization. In fact, aerial views of both, each with its mighty and orderly columns—steel in the one case and stone in the other—bear a striking resemblance." As the child of an oil geologist and a well-traveled teacher, Wanda readily drew the parallels.[36]

She also had some surprising news for her American readers. Abadan was not only in reasonably good shape, but its senior managers were prepared to accept a degree of foreign management "provided Iran shares in that management." Antiforeign feeling, refinery chief Reza Fallah told her, had ebbed in recent months, and workers understood that Iran did not have enough expertise to run the refinery complex at full power. Although they preferred American specialists, some British could be included, Fallah said, "provided they work on [a] new basis, not as in [the] past." He then explained exactly how many foreign technicians he needed for certain levels of refinery output, what he had done to maintain domestic supplies, and how Abadan's situation was totally misunderstood abroad. But Fallah readily acknowledged that his team had been "only partially successful" in combating corrosion since the refinery was hard to maintain with limited operations and financial resources. He said he had "no quarrel" with Hoover's preliminary estimate of how much would be needed to rehabilitate the facilities.[37]

This was not the Abadan that Wanda had anticipated, based on reports she had read in the Western press. Instead of seeing hedonists preoccupied with limousines and motorboats, she found managers who worried about a serious shortage of decent housing for their workers. Expecting chaos, she instead found "no sign of neglect." The refinery was being well cared for: "Everything is spic and span" in the sprawling, two-square-mile complex, she wrote. In the formerly British-only housing area, the lawns were trimmed, and the garden beds weeded. Safety precautions were "exceptional, even by the most rigid Western standards." Cigarettes and lighters, this chain-smoker discovered to her dismay, were confiscated at the refinery gates "to guard against any involuntary slips."[38]

Most of the Iranian staff members she met were educated, cultured, and friendly. The managers were engineers with advanced degrees, doctorates in some cases, from British universities, including Fallah, who would cause considerable difficulties for her a decade later. Despite some understandable grudges against Anglo-Iranian Oil, she wrote, "there is much less recrimination or bitterness than one might have supposed from earlier reports." Operating the enormous refinery even at 5 percent of capacity, given the circumstances, proved their competence, she wrote. These men were gaining valuable firsthand experience. They were the most realistic of all the Iranians she had encountered, and the first to acknowledge Iran's need of assistance.[39]

Foreign involvement, she warned her American readers, would have to be different from what it was in years past. The majors had to recognize that technical and financial problems were secondary to the conflict's human and political dimensions. A tactful approach to Iranians, "a highly cultured, proud and sensitive people whose attitudes and psychology are in many ways alien to the more materialistic West," required "considerably more insight than many Western oil men may realize." Although she criticized the Anglo-Iranian Oil Co. only obliquely, she challenged both American and British executives to look at Iranian managers and workers in a new way. If these oilmen wanted long-term access to Iran's vast oil supplies, they would have to share power.

Wanda spelled out several things she believed the companies must do. First, they should provide decent, low-cost housing to thousands of refinery workers, which as Fallah told her, would "do more than anything else to remove dissatisfaction." The refinery should be run in a "commercially sound" way, she wrote. The oil companies should not only institute a program to "Iranize" the operating staff but also keep as many Iranians as possible in leadership positions. "Supervision of the Iranian staff should also be left in the hands of competent Iranians," she added. "Promotions, demotions or transfers to other posts must be handled not only justly—but demonstrably justly." Employment of unskilled foreign workers and a reshuffling of "senior grade" housing should be avoided or at least handled with "the greatest tact." These actions would "offset anti-foreign propaganda from both ultra-nationalists and Communists."[40] Wanda's step-by-step prescriptions were typical of her educational efforts, whether she was trying to teach Iranians or the majors. She had no qualms about thrusting herself into this mediation role. If Washington political columnists could issue advice on domestic and foreign policy problems, why should she not tell the oil companies how to improve their situation in Iran?

Once she discovered that many Iranian government and NIOC officials were reasonable, she empathized with their predicament. She was an outsider herself, both as a professional woman in the male-dominated oil world and as a naturalized American citizen. Iran's efforts to free itself from foreign domination resonated with this granddaughter of a Slovak nationalist. Although she came to Iran as a reporter mostly concerned about the implications of the oil crisis for her American business audience, she left with a much more sympathetic view of the country and the conflict.[41]

Most Iranian oil officials were impressed by her unusual approach. Her stories were especially appreciated, wrote one senior Abadan manager, S. K. Kazerooni, because accurate information about Iran "is, to a large extent, denied us." He said that he admired her "personal charm which made discussions with you so easy whilst such talks have been known to be quite difficult with others who enjoy your status" as a Western news reporter. Another NIOC official who later became a World Bank official, A. Gosem Kheradjan, wrote in a letter to Wanda that her articles "represent a fair picture of the situation." He believed many foreign "newspapermen fail to do their job properly because they aspire to do too much"; while visiting for just a few days, they think they can "cover a few thousand years of history, tackle all the religious and moral problems and sum up the whole political and economic situation." Instead, Wanda seemed "like a scientist in your exactitude and your limiting yourself to the subject on hand." She ought to become Abadan's "public relations officer in chief," he teased, "because with your neat beautiful figure and cheeky bonnet moving amongst us and preaching the gospel of brotherly love, you would melt the heart even of the most staunch anti-foreign nationalist." M. A. Hirmand, an NIOC official in the nearby city of Khorramshahr, wrote Wanda to praise "the supernatural wisdom owned by such a clever writer."[42] To these Iranians, Wanda so defied familiar norms, so resisted categorization—as a Westerner, a reporter, and a woman—that she seemed almost otherworldly.

She did not win over everyone, however. Some Iranians—though it is unclear who—accused her of being a foreign agent. Wanda ultimately left the country in haste. Many years later, in recounting what happened, she told a gripping, though probably embellished, story of this unexpected departure. Anxious NIOC officials in Abadan, suddenly worried about these "foreign agent" charges, decided to send her by boat from Khorramshahr across the shark-infested Shatt-al-'Arab waterway to the port of Basra in Iraq. Because of tensions between Iran and Iraq, she was dropped off on a deserted island near the Iraqi coast. "Somehow I waved down an Iraqi patrol boat," she recalled. "You can imagine the reaction. A lone woman with an American passport stranded on an island." She was finally picked up and taken to Basra. There, she would have been locked up had she not been rescued by an attentive British attaché.[43]

Although the exact circumstances and timing are not clear, since she told several versions of this incident to colleagues, Wanda stayed only briefly in Basra, although she still managed to file a story. She then arrived in Kuwait sooner than expected, according to L. T. Jordan, general manager for Kuwait Oil Co., a U.S.–British joint venture. He was upset, he later wrote to her, that she had arrived and departed while he was out of town. One NIOC official, Jawad Khalili, later wrote to thank her for her visit to Iran. He teased her for being "taken for an 'agent' during your recent trip," and regretted that she would not soon return, as she apparently had planned. She had, he said, "successfully managed to be an 'agent' of the consortium and the N.I.O.C. at the same time."[44]

BITTER FACTS

However brief, these visits to Iraq and Kuwait stoked Wanda's nascent skepticism about the majors' heavy-handed policies in the Persian Gulf region. While in Basra, Wanda met an old acquaintance of her father, Sami Nasr, one of the few Arab managers employed by any subsidiary of a Western oil company. Nasr was chief geologist for Basra Petroleum, an affiliate of the Iraq Petroleum Co. (IPC), the European-American joint venture that owned the sole concession to explore and produce Iraq's oil. Nasr liked the story that Wanda later filed, based on her discussions with him, about the effect of Iran's new oil exports on Iraq. He was impressed by how quickly she grasped the issues in one day, and equally impressed with what she expressed privately to him—her respect for many of Iran's oilmen and her concern about the majors' insensitivity to the emergence of nationalism in the Middle East. "The bitter facts still remain that we, as oil operators, have not succeeded in endearing ourselves to the people" of the region, he wrote to her. "There must be something basically wrong. The people of the Middle East are not an ungrateful lot." Even after operating for decades in the Persian Gulf, the "oil brass" failed to understand even the most basic problems stemming from the Western companies' exclusive control of the region's oil.[45]

In Kuwait, Wanda met someone even more forthright about the oil brass than Nasr was—James MacPherson, another friend of her father's and a former chief of field operations in Saudi Arabia who had become chief geologist for the independent U.S. company Aminoil in 1954. The industry's "big boys," he railed, knew little about the Middle East and cared only about profits. Wanda must have responded in kind, because MacPherson decided she was different from most reporters. "I admire the way you do business," he wrote her two months later. "You don't quote people when I ask you not to and that's the way I would like to deal with you. Ask me for any information," he added, "and if I have it I would be glad to pass it along." He would become one of Wanda's favorite sources.[46]

Wanda then ventured into Saudi Arabia, a country closed to foreigners. Most Western reporters could not get in except as a guest of Aramco, a joint venture of four U.S. majors: Texaco, Socal, Jersey, and Mobil. Although privileged with a rare invitation, Wanda filed just one main story because Aramco only invited her to the eastern oil fields for a week. Although it was an update on a 1948 discovery, it was an important story because Aramco officials were willing to talk about drilling results in detail for the first time. "Americans are uncovering an oil pool here," she reported from Uthmaniyah, Saudi Arabia, "that measures, not in miles, but degrees of latitude." The Ghawar oil field "is the most fabulous thing that's ever been hit in the world—even in the Middle

East where everything in oil is huge." The new field, she learned, "holds at least as much oil as all the presently proven United States' reserves of 28 billion barrels." Although initially developed in a "dull, rather routine way," three large discoveries now appeared to be part of one huge underground lake of oil. After the Aramco managers pulled out their charts and plotted their progress, they took her on a trek across part of the field to show their final drilling efforts to confirm its size. "But how does it look?" she asked one senior Aramco official, trying to get him to be more precise. "The cat-that-swallowed-the-canary expression on his face," she concluded, "was eloquent answer."[47]

Wanda returned to New York that March of 1954, her bags more than full. She lugged two copper coffeepots—bargains from a Saudi open-air market—and plenty of material to write several more analyses of the Middle East's supply predicament. Upon arrival, Wanda immediately made the rounds of her oil-company friends. She was a big hit, Jaqua recalled, "swamped with requests" for meetings. Even the Central Intelligence Agency wanted to talk with her. In the 1950s, it was not unusual for the CIA to informally interview reporters, academics, and businesspeople traveling to restricted regions, including well-known journalists such as Walter Lippmann and Joseph Alsop. CIA Director Allen Dulles himself cultivated contacts with reporters, including Wanda on occasion.[48] So when a CIA officer asked her if she would informally talk about her trip, she agreed. The agent told her to wait at a certain street corner at a certain time. He and a colleague picked her up in a nondescript car and drove around the city while they plied her with questions about the people she had seen and the places she had been. She was not impressed. These men did not know much about oil or the Middle East. Jaqua recalled that Wanda found the whole scene "pretty funny."[49]

Wanda was blunt, no matter whom she spoke with. Whether talking with CIA officials or company executives, she made them listen to the "bitter facts" she had learned from Iranians such as Naficy and Ala and from oilmen such as Nasr and MacPherson. If they had any common sense and considered the implications of their actions, she said, the majors would find ways to share power and to make more local people managers—and the U.S. government should insist on this change.

However powerful they might be, the companies should not hide behind legal contracts. They should look down the road a bit. Their success in frustrating Iran's nationalization bid and finding vast new
reserves in Kuwait and Saudi Arabia should not blind them to the political implications of their actions. Oil nationalism was bound to shake
up the Middle East.[50]

ONLY WORK MATTERED

Emboldened by her trip, Wanda was ready for a new challenge. The
Journal of Commerce had given her considerable free rein, but she felt
that her dynamic, if not quite elegant, style got bogged down on the paper's staid pages. New York harbor fuel prices and Oklahoma drilling no
longer fascinated her. She wanted to focus on international oil news,
but her editors would not let her do that, so she looked for another job.
Initially, she pinned her hopes on the *New York Times*. She had been
"running circles" around its oil reporter with her coverage, Jaqua recalled. "She loved to pick apart his work—just tear it up—because it
was so bad. [He] had none of Wanda's insight or contacts or skill." But
the *New York Times* was not hiring women, and certainly not female
business reporters. Aside from the handful who wrote for the women's
pages, fewer than half a dozen female reporters worked for the *New York
Times* in 1954, none of them in business news.[51]

A larger problem, however, was that Wanda now had personal issues
to deal with. Her lengthy trip to the Middle East had not been good for
her marriage. "She had jumped at the chance to go," Jaqua remembered. "She could hardly wait to get out there." Although she wrote him
occasional letters, it was hard to have her away so long, especially during
the holidays. Though Jaqua admired her drive, what he really wanted,
Wanda told friends and colleagues, was what most of his peers had—a
wife at home, raising children, cooking dinner, and making life comfortable. Her obsession with work may have been all right in their early
years together—they had now been married nine years—but by the
time she left for Iran in late 1953, it was not.

Wanda's absence during that winter must have forced Jaqua, thirty-
six years old, to reevaluate their situation. By then a partner at Sullivan

& Cromwell, he was more than ready for a family. When asked whether Wanda was unable to bear children, Jaqua simply answered, "She did not want them." Even daily domestic routines became irritants in their marriage. She hated the kitchen. When she returned from the Middle East, she went back to her usual work routine—staying out late. Thanks to Jaqua's new status and income, the couple moved uptown to an expensive apartment overlooking Central Park, which Wanda enjoyed decorating. But neither she nor Jaqua was happy. "Wanda was very obsessive and always running off to do things and leaving Jack," recalled her friend Eleanor Schwartz. She saw Wanda flirt with one of the *Journal of Commerce*'s editors, and Jaqua developed an interest in a secretary at his office.[52]

By then, Wanda rarely brought her husband to parties, though she occasionally had in the first years of their marriage. "Wanda worked night and day," Walter Levy recalled. "No one even knew she had a husband." Farmanfarmaian remembered meeting Jaqua at a dinner in their Central Park apartment—"an extremely handsome chap, but subdued, like a very obedient servant." Wanda ran the show. The Iranian aristocrat was not used to "this breed of men who let their women dominate." Nor were Eric Drake and his wife. Drake was Anglo-Iranian Oil's director in Iran until the company was nationalized in 1951; then he was assigned to New York, and he later became chairman of British Petroleum. When Wanda invited Drake and his wife to dinner shortly after their arrival, they were surprised that her husband cooked and served the dinner. "In those days, it was very unusual," Drake said years later. "She was tough. She wasn't domestic at all." He and his wife liked Wanda but sympathized with Jaqua because they thought Wanda treated him badly. Jaqua appeared willing to forge a less traditional marriage, but even those attempts foundered.[53]

Wanda also had professional problems with Jaqua. At one point in the early 1950s, a New York court subpoenaed her in an oil price-fixing case. As a reporter, she knew she could not reveal information on her sources, but she was worried and asked her lawyer husband for advice. Surrender your notes, he told her. "Of course I'd say that," Jaqua later recalled. But Wanda was incensed, devastated. How could he not see that her reputation as a reporter was at stake, even her career? His failure to see her point of view as a journalist infuriated her. The next day, she walked into court with an overnight bag in case she was jailed, but she was dismissed without being cited for contempt. The incident, her friend Margaret Clarke recalled, left her embittered. Their marriage continued to deteriorate.

Wanda began to acknowledge to a few friends that work was incompatible with marriage. But perhaps it was that her temperament was not suited to marriage. She was not a nurturing person. Her globe-trotting childhood prepared her for her career but not for a long-term commitment, especially not for one decided in haste during the war. Friends such as Eleanor Schwartz said that she carefully camouflaged her marital problems. [54]

Despite the strains between Wanda and Jaqua, nothing got resolved at that point. After all the success she had from her Middle Eastern trip, Wanda instead threw herself into looking for another job. Disappointed that the New York Times would not even consider hiring her, Wanda continued to search, and by summer she had found a good alternative. A team at McGraw-Hill, the country's leading business publisher, persuaded her to help launch a new oil magazine—Petroleum Week—to rival the lucrative but stodgy Oil & Gas Journal. Georgia Macris, foreign editor for McGraw-Hill's news agency, "gushed forth about [their] coup in hiring the 'Great Wanda' away from the Journal of Commerce," recalled C. B. Squire, a colleague. Wanda could now set her own terms: she would be international editor, with carte blanche to write when she pleased and as she pleased, whether news stories or signed commentaries, from New York or abroad. No one would be allowed to touch her copy. [55]

SUBTERRANEAN POWER SHIFT

In August 1954, when Iran and the consortium of majors finally signed an agreement resolving the three-year nationalization crisis, most of the American press declared it to be an outright win for the West. Business Week, for instance, billed the accord a "resounding victory for the oil companies," declaring that Iran's "Oriental drama" had left the country bankrupt. "Little is left," the magazine declared, "of weepy, wily ex-

Premier Mossadegh's grandiose dreams of going it alone in the world oil market." The majors' boycott had succeeded. By retaining a 40 percent stake in the consortium, Anglo-Iranian Oil, renamed British Petroleum in 1954, had regained a "controlling position" in Iran. *Business Week* concluded that it was a "good deal no matter how you look at it."[56]

In her final news story for the *Journal of Commerce*, however, Wanda had a different take: she described the deal as equitable to both sides. But once she left the paper and before *Petroleum Week* started up with its first issue in 1955, she struck out more boldly with a freelance article in the popular magazine *Collier's*. Do not pigeonhole Iran, she warned in the lengthy piece, titled "Master Stroke in Iran." Its oil saga was not over. Americans should not "slip back to the old, comfortable view of the country as merely the home of handsome Persian cats and carpets." For Iran, oil nationalization "remains a semi-mystical symbol of independence from foreign intervention—Iran's own Boston Tea Party. Oil never was the primary issue in Iran (any more than tea was in Boston)." Many Iranians "are convinced that their continued freedom still depends, as it has for centuries, on playing off one foreigner against another." Mossadegh, though in prison, was still "something of a national hero—a pajamaed George Washington of modern Persia," she added, referring to the former premier's predilection for receiving official visitors in casual attire. Iran's "well-meaning but often indecisive" shah had approved the oil pact, but it would be meaningless if Iranians did not quickly see some tangible results. The West had benefited from Moscow's clumsy public protest when it had warned Tehran against signing the oil pact, but the West should take note of Iran's reaction, which she summarized this way: "Nothing makes a Persian angrier than open interference in his country's affairs."[57]

Although she worried that Iranian nationalists might be "more inclined to throw bricks than to build with them," Wanda expressed sympathy for their concerns. "Iran is a land which is often difficult for Western realists to understand." Its people "are famed for their hospitality and great charm but at the same time notorious for their volatility and mysticism. They value their poets more than their plumbing, and talk of their glorious past more than their future," she wrote. "Even the illiterate common laborer can often quote the classics."

The consortium agreement was indeed historic, she noted, but not only for the obvious reason that it was the largest single business transaction on record. It was also a "unique example of big-scale business diplomacy" and a victory in the West's struggle to contain Soviet Russia. (The British and U.S. intelligence agencies' role in branding Mossadegh a communist sympathizer was not public knowledge at the time.) But there was something more that the press did not report, she wrote, something the majors did not want brought to anyone's attention: under pressure from Washington, the companies had conceded for the first time that the oil-producing country, not the concession-holder, legally owned the subsurface oil. Forced to abandon the typical Middle Eastern concession, the majors had to settle for a contract. Although they would still have effective control over Iran's oil operations, the Western companies had set a "dangerous precedent" for themselves, Wanda explained, by forgoing title rights with this "formal bow to Iran's sense of ownership." Therefore, Iran's legal victory was also historic.[58]

Jersey's Howard Page, she went on to explain, disagreed with her assessment. Known as the best brains in the business, this Jersey director had taken over as the consortium's chief negotiator in the final sessions. Wanda admired him, but in this case she took issue with his position, as she would at several critical moments in later years. Page, she wrote, "pooh-poohs fears" of an ill-advised precedent. What was the difference, she quoted him saying, "between owning a car and just borrowing it for forty years?" Everybody knew that Iran would not be able to tell the consortium what to do until 1994.[59]

Wanda answered him with a direct rebuttal. Iran "can legitimately regard the pact as a major achievement," she argued, given the financial clout of the majors and the glut of oil on the market, since the consortium also committed to quickly ramp up production. What's more, Iran would pay only a minor amount in compensation for Anglo-Iranian Oil's installations and not a penny for the company's loss of earnings from the three-year shutdown or the remaining four decades of its original concession. And Iran's most significant achievement, Wanda argued, was that the agreement had the potential to change the balance of power in the Middle East—not between the cold war powers of West and East, but between the majors and the oil countries. Just as the

Venezuelan oil nationalists had claimed ownership in 1948 and insisted on a fifty-fifty split of profits, so had the Iranians won a formal acknowledgment that their country had title to the oil beneath them. Foreign petroleum companies, at least in Iran, were no longer owners; they were just contractors.[60]

Wanda was not pleased with the published version of the article. It was not all she had wanted to say, she wrote to friends and sources. As she told Herbert Hoover, the chief U.S. negotiator, the article had been rewritten so much "by a 'consortium' of Collier's editors that I hardly recognize it." She knew Jersey's Howard Page did not like it. She also complained to Hussein Ala, the former prime minister, that Collier's had cut a long section on Iran, whereas everything she wrote about Hoover was kept intact. To Denis Wright, she was even more explicit about her dismay: "I am not sure the Iranians will like my description of them in this final version—though my original story which dealt primarily with them (before all the oil negotiations were added) was much more sympathetic." Nevertheless, with the Collier's article, she had reached a broader audience than through the Journal of Commerce. It even prompted Allen Dulles to write and invite her to meet with him in Washington.[61]

By this point, Wanda was more than ready to devote herself to international oil coverage for McGraw-Hill. She was now convinced that the future of the oil industry would be determined by the massive reservoirs of the Persian Gulf region. West Texas was old news. American fields could no longer satisfy the world's insatiable appetite for oil. To meet this surge in demand, the majors were concentrating on the Middle East, with the larger independent companies sure to follow. That's where the big news would come from, and that's what she wanted to cover.

With the Iran consortium agreement of 1954, many American diplomats and oil executives congratulated themselves for putting an end to a troublesome crisis. But to Wanda, this was not the end of a crisis, it was the beginning of a new era. Mossadegh's abortive effort to nationalize Iran's oil industry looked like a failure to many Western observers, but his willingness to defy the British Empire for more than two years galvanized emerging nationalists in the region, and Wanda was paying attention. Through Iran's efforts, nationalists discovered a latent power—a power that could not yet be fully unleashed, but a power nonetheless.

The idea that oil equaled power had been the subtext of many of Wanda's interviews during her trip, including one in Tehran that she described in detail: "Mark my words well," a highly placed Iranian told her while a servant served ceremonial glasses of tea in silver filigreed holders. "The oil operations will remain the central political theme of Iran for a long time to come. Every politician who is in power, and every politician who wishes to come to power, will use the oil issue to his own ends."[62] For Wanda, this power struggle over oil fascinated more than anything else.

4.

Savoring the

Bedouin Brew

London in November 1956 was remarkably cold, Wanda found as she stopped off there for two weeks. She was on her way to the Middle East for what turned out to be a trailblazing reporting trip of nearly five months, this time for *Petroleum Week.* She would soon develop a theme that she had begun to explore on her 1954 trip: Americans had to learn more about the Middle East. Concerned by the ignorance about this region among oil executives and Washington officials alike, she set out to give them context and texture, human-interest stories and the latest drilling news, interviews with Middle Eastern leaders and analyses from Western engineers.

When she arrived in London, though, she worried whether she could obtain the visas she needed for even a two-week stint. Visas were hard to get even under normal circumstances, and these were not normal days in either Britain or the Middle East. Her high-level contacts who had promised to help were preoccupied with a national emergency: Britain was running short of oil thanks to a showdown with Egypt over the Suez Canal. London's weather was unusually cold, and, because of an extreme fuel shortage, there was almost no heat.[1]

Earlier that month, Britain, France, and Israel had joined forces to invade Egypt and retake the Suez Canal, which Egypt's charismatic leader, Gamal Abdel Nasser, had nationalized in July. Although they had retreated from many of their colonial holdings around the world, Britain and France would not give up the canal without a fight. The waterway was a critical shortcut to Persian Gulf oil supplies and former colonies in Asia. By seizing the canal, Nasser had struck at a very tangible symbol of nineteenth-century colonialism. Once attacked by superior firepower, Egyptian forces retreated, but they sank enough ships and barges to make canal transit impossible. Arab sympathizers then sabotaged pipelines in Syria and Kuwait, and Saudi Arabia, in a show of solidarity, imposed an oil embargo on Britain and France.

Making matters worse for his allies, Eisenhower prohibited American oil producers from alleviating Europe's shortage with extra output. Although Washington had helped provoke the canal's nationalization by turning down promised funds to build the Aswan Dam, Eisenhower was furious that Britain and France had resorted to force—especially without consulting him—rather than compromise. As a result, by mid-November the British were turning down radiator thermostats and rationing gasoline. What's more, Eisenhower refused to rescue Britain from a run on its already low gold reserves. Faced with American opposition and outright condemnation from the United Nations, the British, French, and Israeli forces withdrew in December. The British, in particular, felt humiliated.[2]

Wanda's London hotel room was frigid, she wrote to her *Petroleum Week* colleagues. So was the executive dining suite of Royal Dutch/Shell, the world's second largest oil company. Invited to lunch by two Shell directors, John Loudon and Sir Francis Hopwood, Wanda "kiddingly asked Sir Francis whether he had purposely had the heat turned down extra low" in honor of a visiting American. Even more surprising was when the Ministry of Fuel and Power's undersecretary, Keith Stock, an old friend, apologized for not driving her back to her hotel after dinner. "He explained, somewhat embarrassed, that he didn't have enough gasoline!"[3] But in describing Britain's painful predicament, Wanda chose not to report these incidents publicly. Instead, she quoted a woman standing in line for a bus: "Those Americans have got all that oil but won't let us buy it. It's a lot like withholding blood plasma from someone because he's been a bad boy. My husband may lose his job. But a lot those Americans care!"[4]

News of Nasser filled the pages of the Western press during those months, but once Wanda reached Saudi Arabia, she sent her editors a commentary with a different take. As far as the geopolitics of oil were concerned, "the No.1 man to watch in the Middle East" was not Nasser, but Abdullah Tariki, an unknown Bedouin who had graduated from the University of Texas with a degree in geology. As the first director of Saudi Arabia's new Office of Petroleum and Minerals Affairs, this thirty-seven-year-old was a "young man with a mission—a force to be reckoned with." Tariki wanted to bring about fundamental changes. Decisions about Saudi oil should be made by Saudis in Riyadh or Jeddah, he told her, not by Americans in New York. He wanted a fairer distribution of profits—and had the numbers to back his position. He also wanted Middle Eastern oil countries to adopt joint policies. She did not quote him saying anything about nationalization or nationalism, but the implications of Wanda's column were clear. Here was a spokesman for a "new generation" of educated Arabs who was putting the international oil companies and their governments on notice.[5]

A PRIMA DONNA WITH A PLATFORM

The elites of the oil world took note, but many of them scoffed. Tariki was a minor bureaucrat, they argued, not a member of the royal family. Wanda, on the other hand, was already a force to be reckoned with, someone who could cause trouble. Over the next five months, she would write what turned out to be an award-winning series of eighteen columns about Middle East oil politics, personalities, and production—more than the industry had ever seen from one journalist. And though the oil executives eagerly read her reports, they were sometimes stung by the results. She gave extensive coverage to their burgeoning operations, but also focused on the needs and aspirations of the people of the Persian Gulf region. In the process, Wanda raised serious doubts about a fundamental precept of the international petroleum industry: that the oil deposits the companies found abroad through their own ingenuity, investment, and hard work belonged to them alone.

Petroleum Week's editor highlighted her trip in a press release by explaining that, as the "petroleum industry's only woman feature writer,
Wanda does a man-sized job." Oil executives "regard her not as 'just a
woman editor' but as a well-informed expert. Wherever she goes they
call her 'Wanda' and know her as . . . somebody who has a talent for digging in the right places to get stories for oil men. She doesn't write a
woman's page."[6]

For its part, *Petroleum Week* quickly established a reputation for being a substantive paper. Though oil trade publications emanated from
New York, London, and the Midwest, their reporting "was wide but not
deep" because the majors were "almost hermetically sealed" from the
public, recalled Massachusetts Institute of Technology economist M. A.
Adelman. "People today don't realize how vastly different it was, how
few controls there were on the international oil industry." Through interlocking partnerships with the Iraq Petroleum Co., Aramco in Saudi
Arabia, and the Iran consortium, the Seven Sisters still controlled almost all of the international business in the mid-1950s, and they kept it
"all a mystery," Adelman said. The American oil press generally lived off
of press releases and "lacked independence," particularly on the world
scene. Senior oil executives "simply did not want or need to talk with reporters—except Wanda." She changed the whole picture; her investigative style of reporting was "such a breakthrough." With the appearance
of *Petroleum Week* in mid-1955, Adelman said, "I knew the difference
immediately."[7]

For McGraw-Hill, the magazine was just one more "book" in its already dominant portfolio of business journals, but one designed to tap
into the wealthy pool of advertisers that had made *Oil & Gas Journal*
so lucrative. For its nearly thirty magazines, including the flagship,
Business Week, and several daily and weekly newsletters on domestic oil
issues, McGraw-Hill could draw on its international business news service, World News, to supplement the *Petroleum Week* staff.[8] Domestic
coverage for the new magazine, Adelman recalled, came mostly from
"good old boys from the Oil Patch," but the international team that
Wanda helped assemble proved "first-rate."

"When she first hired me to work on *Petroleum Week*," remembered
Georgia Macris, a World News staff member since 1947, "I knew noth-

ing about oil. She taught me well." Wanda trained reporters to write for "some little oil man sitting in Oklahoma or Colorado" as well as for the president of Jersey. She was a "relentless taskmaster, ... very dynamic but also very patient." A Radcliffe graduate from a Greek immigrant family who became foreign news editor at *Petroleum Week*, Macris worked with Wanda longer than any other journalist did. Wanda and Macris also cultivated three talented correspondents: John Pearson in Caracas, later a *Business Week* editor; Helen Avati in Paris, known for her scoops; and Onnic Marashian in Beirut, later the longtime editor of McGraw-Hill's *Platt's Oilgram News*.[9]

McGraw-Hill's newly assembled passel of oil publications settled in at the company's Forty-Second Street headquarters, known as the Green Pagoda for its distinctive shape. Staff members were grouped in a large, central newsroom, and editors occupied window offices. Wanda, however, spent little time there. Even when not traveling, she was usually out seeing people, then dashing in right on deadline, copy in hand. "She worked around the clock, it seemed," recalled a colleague, "Chick" Squire. Her competitive, prima donna personality scared many on the staff. She "seemed incapable of small talk," Squire said, so he remembered his surprise at seeing her "unbend a little" once at a staff party.[10]

What's more, she was not a facile writer. She was given to pacing the hallways, cigarette in hand, fuming. "Wanda was a perfectionist. She would agonize over her writing, sometimes taking all night long to write her columns," Macris remembered. "Other nights, she would corral me into staying up until 2 A.M. going over every single sentence, every single word." *Petroleum Week*'s managing editor, Dick Machol, "used to say that when Wanda was writing a story, it was worse than a woman having a baby," a friend, Harriet Costikyan, recalled. Wanda herself admitted to this frustration: "I write for a living," she once told another friend, "and I loathe writing." With her talent for gathering off-the-record information, she constantly had to decide what she could use explicitly and what she must allude to implicitly. Her most astute readers, her colleagues said, knew how to read between the lines of her carefully crafted articles. Her editor and publisher knew they had to handle her, Macris said, "with kid gloves."[11]

LOBBING BOMBSHELLS

Starting with *Petroleum Week*'s first issues in 1955, Wanda honed in on the Middle East. The four leading oil-producing countries—Kuwait, Saudi Arabia, Iran, and Iraq—had different production histories and concession deals but overlapping interests. Kuwait, still a British protectorate, boasted the highest output thanks to a British Petroleum–led joint venture with Gulf Oil called the Kuwait Oil Co. Iran's production was beginning to revive under the direction of the newly created Iran consortium, with 40 percent controlled by BP (the former Anglo-Iranian Oil Co.) and most of the rest split between the other six majors (Shell, Jersey, Mobil, Texaco, Socal, and Gulf) and France's Companie Française des Pétroles (CFP, today known as Total). The Iraq Petroleum Co. (IPC), which had Iraq's sole oil concession, consisted mainly of BP, Royal Dutch/Shell, two American majors (Jersey and Mobil), and France's CFP. But Wanda's top priority was the only all-American joint venture in the Middle East, the Aramco partnership between Socal, Texaco, Jersey, and Mobil.

Aramco, the Arabian American Oil Co., was the largest single investment of private American capital abroad. A dozen years after Socal drilled its first gusher in 1938, Aramco made Saudi Arabia a leading oil exporting country. A first taste of wealth whetted the royal family's appetite for more oil dollars. As production grew, however, corruption flourished, and by 1950 the kingdom was broke. The Saudi king, Abdul Aziz ibn Saud, founder of the modern Saudi state, demanded what the Venezuelans had secured in 1943: a fifty-fifty split in profits. Aramco's 1949 profits were nearly three times greater than its royalty payments to the kingdom, and even the U.S. government got more in taxes than the Saudis got in royalties. With a fifty-fifty split, the Saudis could nearly double their yearly income, and the deal would not be a big drain on Aramco because of a U.S. law protecting international companies from double taxation. The big loser on this deal would be the U.S. Treasury because Aramco could deduct from U.S. taxes all its payments to the king.[12]

That was fine with Washington. With the onset of the Korean War, the Middle East became critical as a bulwark against communism. The Truman administration wanted to maintain friendly relations with Ibn

Saud, contain radicalism, and ensure a steady supply of petroleum. Thus, with Washington's assent, the Aramco companies agreed to the fifty-fifty split.[13]

One of Wanda's best Middle East sources at that time was her father's friend James MacPherson, whom she had met during her trip to Kuwait in 1954. By the middle of 1955, MacPherson was about to defect from the industry by becoming a consultant to the Saudi king and personal advisor to Abdullah Tariki. A small, wiry, almost gnome-like Scotsman with a slender mustache, "Mr. Mac," as his colleagues called him, was in charge of Aramco's field operations in the late 1940s, when Saudi production grew dramatically. He was popular with the field staff, even though, as one reporter put it, "his language can be as blistering as it is eloquent." Frustrated with the increasingly bureaucratic policies and politics of Aramco's parent companies, he got upset when the four-company board ordered Aramco to slow production because of the growing glut of oil on world markets. So MacPherson quit in 1949, a U.S. diplomat wrote, out of "disgust" for the board's insistence on "balance sheets and profit before all else." He became chief geologist for the independent U.S. company Aminoil, which made huge discoveries in Kuwait, to the dismay of his former bosses. By 1955, as he began working as advisor to the Saudi king, MacPherson found in Wanda an eager listener to his sympathetic views of Arab nationalists and contempt for the major oil companies.[14]

He proved an excellent source of information. Before leaving Kuwait for Beirut, where he would set up an office, MacPherson gave his favorite reporter some "dope" on condition that she be "cagey" in how she used it. "You are the only one getting it," he wrote Wanda in mid-1955, "so protect me." Kuwait Oil (KOC), the majors' joint venture, he told her, was renegotiating its profit-sharing agreement with the ruler of Kuwait "in a great atmosphere of secrecy." There "is dynamite in this," MacPherson warned. "It will probably be denied. KOC personnel are all pledged to secrecy against leaks." He gave Wanda the details because "you know my idea on secret deals. They always cause trouble." He did not mention, however, what she would soon find out on her own: Jersey and Mobil, two of the four Aramco partners, were also involved.[15]

Wanda used the tip well. She got confirmation of the deal—and more. Kuwait, as it turned out, was not only revising its agreement with Kuwait Oil but also completing secret negotiations with the main buyers of Kuwait Oil's production: Shell, Jersey, and Mobil. The deal would give a surprising new twist to the fifty-fifty profit-sharing deal that had become the norm in Middle East oil concessions: even though Shell, Jersey, and Mobil were just buyers, not producers, Kuwait insisted that they pay, in effect, a sales tax that had not previously been levied. Wanda explained the intricate details in her column, point by point, and to top it off, she coyly noted that none of the companies seemed to agree on the deal's implications, with some calling it "reasonable" and others "a dangerous precedent."[16]

By breaking this news and exposing disagreements among the majors, Wanda showed in the first official issue of *Petroleum Week* in July 1955 that she was playing hardball. MacPherson loved the results. He sent her a copy of the letter he wrote to the Saudi finance minister that included a United Press report on Wanda's scoop and MacPherson's own suggestion that the minister immediately subscribe to *Petroleum Week*. But she had to be careful, he warned in a scribbled note. In the future, she should mail him letters in a plain envelope marked "Confidential," with no return address.[17]

A few months later, MacPherson asked Wanda to visit him again in Beirut: "Would discuss verbally much of interest that would make interesting articles." Wanda was chagrined. "That's like dangling a bone in front of a dog and then telling him, 'No,'" she responded. She had just returned from several trips within the United States and one to Europe, so it would be awhile before McGraw-Hill would pay for her to go to the Middle East. In the meantime, she explained, her contacts "keep me a bit informed." "No one tells me too much, but everyone tells me a little —so I just talk to a lot of them, and it adds up."

But MacPherson helped direct her toward another bombshell later in 1955. "I've heard the big boys rave on about the 'sanctity of the concession terms,'" MacPherson wrote. To him, the terms were simply unfair, given the profits they made on hidden discounts. It was even more galling because the majors were flush with record earnings and their stock prices were at an all-time high. "Wanda, you and I are in the wrong business," MacPherson wrote. "Looks as if the Ruler [of Kuwait] has—putting it bluntly—been taken for a ride."[18]

A few weeks later, Wanda exposed those discounts. She explained, country by country, how the majors had not really divided actual profits on an equal basis, but used discounts to affiliated companies and other sales allowances to distort the true balance sheet. Since Middle East governments were denied access to the companies' books, this disparity, "not unnaturally perhaps, has raised uncertainty and suspicion," particularly since these officials "have little experience in oil policies and are insecure in their comprehension of the international oil trade."

What was at stake here, she argued, was far more than just money. While "the Arabs have more than mastered the fine art of squeezing the goose that lays the golden barrels," the rulers "want not only more profits, but profits tied to some kind of public pricing—publicly explainable and understandable." Stable relationships among companies, governments, and the public, she wrote in her column, "depend as much on acknowledged fairness as on compliance with laws and contracts." Her stance mirrored MacPherson's. His indignation made her more attuned to the oil nationalists. And for companies posting big profits precisely because they set prices and kept all transactions secret—at least in international deals—her argument for more transparency was unsettling.[19]

WHISPERING IN BEIRUT

As Wanda was making final preparations in London in November 1956 for her trip to the Middle East, she received, she later recalled, a disturbing cable "from Beirut strongly suggesting I stop there for awhile first—without quite making it clear why." She did not find out what was wrong until she arrived in Lebanon in mid-December: Abdullah Tariki, the top Saudi oil official, had denied her a visa because he did not like her views on the Suez crisis.[20]

Months earlier, before the Suez conflict began, she had argued in a column that the United States had a "strategic stake" in the Middle East because the Soviet bloc, by supplying arms to Egypt, threatened oil shipments through the Suez Canal. Any cutoff would not only cause shortages in Europe, but destabilize the region so much that the Soviet Union would have "half won the cold war." Press reports about sheikhs with Cadillacs and opulent palaces, although "quite true in some cases," had misrepresented the people of the Middle East. For them, she explained, "oil revenues mean water [and] water means life." Their leap into "contemporary civilization" was so rapid that the West needed to show greater patience and understanding. But when Nasser national-ized the canal in July 1956, Wanda denounced the move, writing that the takeover undermined the entire system of Middle East oil conces-sions and contracts. Gulf Oil's chairman sent the column to Secretary of State Dulles, noting that it was "extremely pertinent."[21]

When Wanda found out in Beirut that her column had jeopardized her invitation to Saudi Arabia, she placed a call to Tariki from her hotel, the St. Georges. At first she fretted as she waited at the bar for him to re-turn the call. Chain-smoking, sipping whiskey, and chatting with other patrons, however, she came to realize that she was not wasting her time. The hotel bar was, in fact, the prime spot in Beirut to pick up news tips. "You can see just about anybody who is anybody in the Middle East" at the St. Georges, she wrote to her colleagues. No wonder Beirut, the "trade and money capital" of the region, was known as the "Eyes and Ears of the Middle East." Within fifteen minutes at the bar, Wanda ran into several Aramco senior executives, two auditors from Price Water-house, the British ambassador, and, of course, Mac MacPherson.[22]

Yet this port city, a small-scale Monte Carlo "with its warm, turquoise Mediterranean, its curving beaches, and its modern buildings rising against the breathtaking backdrop of the snow-capped Lebanese mountains," felt like a ghost town. The Suez crisis had almost brought business to a standstill. British and French tourists were "conspicuous by their absence" as Arab countries clamped down further on foreign visitors. Shipments through the normally busy port of Beirut were "down to a trickle," she reported.[23]

The elegant St. Georges, however, was not quiet. This French art deco hotel, a landmark building on a dramatic little peninsula jutting out into the Mediterranean, seethed with rumors and conspiracies. "Ev-eryone seems to be watching everyone else—like a nightclub full of cats and mice," Wanda wrote to her New York office, not for publication.

"My companions—residents of these parts—kept whispering, 'See that skinny one over there with the waxed mustache? He's so and so, who made millions in the contracting business because his father-in-law was a minister.' 'Notice that fat one just coming in, looking like Farouk II? He's the slipperiest character in all Beirut. Worth millions.' 'That one in the corner, with those other three is a well-known agent for so and so in Syria, and who's backing the Communists so he can . . .'" It was "just the place you had to be" if you were a reporter, said Wanda's colleague Chick Squire, who had worked in Beirut for McGraw-Hill and the local *Daily Star.* "If there was a crisis, I would check in there every day."[24]

Eventually Tariki called her back from Jeddah. "Nasser is God," she remembered him saying. "We're all Arabs and all Nasser people." Unfazed, she retorted that he ought to let her into the country, otherwise she might really become anti-Arab. Tariki backed down and agreed to arrange for her visa. Pleased with her successful bluff, Wanda went off to Mac and Grace MacPherson's home for dinner.[25]

The next day, Onnic Marashian, *Petroleum Week*'s new reporter in Beirut, met Wanda for lunch at the St. Georges. An Armenian with a journalism degree from Cairo's American University, Marashian was, Wanda felt, "one of the best informed, accurate, and thoughtful correspondents in the Middle East." Marashian had worked for the *Christian Science Monitor* in Cairo before opening McGraw-Hill's Beirut office. Although fluent in Arabic, he had to cope with the suspicion most Arabs felt toward non-Muslim Middle Easterners. Even Tariki later warned Wanda about this problem, although he liked Marashian.[26]

For his part, Marashian had to overcome his own suspicions—of assertive women. "My mind-set back then was very Middle Eastern," Marashian recalled. "So of course a woman correspondent was something unbelievable to me." Moreover, "she was going to be my boss." The fact that she was a woman "made me very careful about how I behaved." He knew of her reputation: "From Kuwait to Cairo," Marashian had written to her, "even local people who do not know you personally are familiar with your name and articles." She was generous with him, he recalled, "but I kept my distance from her then because she was so famous."[27]

Despite his admiration for her, Marashian was totally unprepared for what happened when he walked into the St. Georges restaurant with Wanda. Leaders of Iraq's IPC joint venture were dining there together with the British ambassador to Lebanon, and when this group caught sight of Wanda, "they all stood up in deference and paid homage to her," Marashian remembered. "All these big shots were practically on their knees giving her their greetings." He was nonplussed: "I had never seen men acting like that toward a Western woman." These were "people who wouldn't even look at me," though he had been covering oil for the U.S. press for two years. Wanda was gracious, even imperial. She already "had the mentality of a chief executive." After receiving their greetings and chatting briefly, she glided over to a table on the far side of the restaurant, sat down, and ordered a scotch — "my breakfast," she said.[28]

All that Wanda reported to the home office about the lunch was that she had seen "my old friend, Sir Stephen Gibson, head of the Iraq Petroleum Co., with his twinkling, beaming face in odd contrast to the half-dozen solemn-looking IPC officials flanking him, all just back from talks in Baghdad." She had stopped to chat, she told her colleagues, and "Did that start a host of rumors!"[29]

ROYAL TREATMENT IN SAUDI ARABIA

The next day, Wanda got up before dawn for an unusual flight to Saudi Arabia. Because Syria had restricted its air space to all but Russian fighter planes, the Aramco-chartered aircraft had to make a detour through Turkey to get to Dhahran, in eastern Saudi Arabia. She had originally planned to visit Syria, a critical transit country for Aramco and IPC shipments destined for tankers in the Mediterranean, but when the pipelines were sabotaged in November 1956, Wanda told her home office, "my itinerary blew up along with them." It took an hour to refuel in Ankara — "high octane gasoline for the plane and high octane vodka for the passengers" — and she shivered in the unheated airport as she watched a couple dozen Turks on the runway "with old-fashioned little ice-picks, chop, chop, chopping away at the ice by hand!"[30]

Next, the plane stopped in Baghdad for another refueling. Iraqi immigration officials took away the passengers' passports so that they would not be tempted to "skip out of the airport and immigrate," they were told. When the plane was ready, the officials were nowhere to be seen, and the passengers finally found their passports lying in a heap by a door. Airborne over Kuwait, Wanda strained to see the two wells that Beirut Radio said had been sabotaged, but "everything looked quiet and normal." An icy wind was blowing when she arrived in Dhahran at dusk.[31]

The Aramco compound was an American outpost of three thousand employees and family members, who lived in small trailers and prefabricated homes on the edge of the vast desert. Aramco was still very much a frontier organization, but Wanda was struck by the settlement's familiarity. It "looks like a lovely suburb in Houston," she wrote, with Christmas trees in the windows and strings of lights adorning homes with neat, green lawns. There were no hotels, only guesthouses, and Aramco put her up in its finest, the one reserved for visits from the king. Her late-night international dinner was fresh eggs and steak from Australia, fresh vegetables from Eritrea, fresh fruit and salad from Beirut. She slept past noon the next day. Both Aramco board chairman Fred Davies and president Bob Keyes came to greet her and invite her to dinner. She was treated, she said, like a queen.[32]

After her visit in Dhahran and lengthy interviews with Tariki in Jeddah, which led her to naming him the "No. 1 man to watch," King Saud sent permission for her to come to his court in Riyadh. On Christmas Day 1956, she flew into the capital in a tiny chartered plane for what she hoped would be an important interview. From the air, Riyadh looked like a jumble of sun-baked mud structures, barely distinguishable from the tawny brown of the empty desert surrounding it. "There it is, honey," the American pilot said, pointing out the earthen-walled palace as he eased down toward Riyadh's primitive, one-strip airport.[33]

For the royal family, Riyadh was the heart of the kingdom. It was there in 1902 that Saud's father, the young Ibn Saud, and a band of raiders had scaled the walls of the city's citadel and reestablished the rule of the Saudi tribe. Ibn Saud gradually expanded his kingdom to include most of the Arabian peninsula, but it was not until the 1950s that

Riyadh, deep in the desert interior, became more than just a dusty garrison town. Indeed, ten years earlier the only way to reach the capital had been by camel—a twenty-one-day trek from Jeddah.

When Wanda arrived, the city seemed to be a massive construction site, still tasting and smelling of frontier rawness. The king had recently decided to move all government ministries to Riyadh from the Egyptian-style comfort of Jeddah, the Red Sea port city. Newly widened streets, swirling with dust, were carved out of the rabbit-warren density of the old town. Unfinished steel and glass office buildings jutted up from the dirt along still unpaved boulevards. The city's population had doubled in one year to two hundred thousand, but the highly religious mores of the desert prevailed. The day and the clock began at sunset. Despite some Cadillacs and gold watches, it was easy to see that only two decades earlier, dates, camels, sheep, and goats had been the primary source of trade. Few Westerners lived in Riyadh in 1956, and almost none were women.[34]

Even with all her travel in the Middle East since her teen years, Wanda felt odd in Saudi Arabia. Almost everyone she talked with, whether Western or Arab, seemed amazed to be meeting a female oil reporter. "And since women in this part of the world don't go around without having their heads and faces covered," she wrote her editor, "I feel doubly exposed." Incidentally, she added, "forget about everything you ever saw in the movies about the Arabian Nights. Instead of glamorous chiffon veils, the women wear either ugly black leather masks (Bedouin women) or a black stocking-like covering over the whole face. All most unattractive and shapeless."[35] At that time, non-Muslim women did not have to follow Muslim law. "Western professional women were not looked upon as women, but as men" by the Saudis, says Middle East expert Phebe Marr. They were not seen as a threat to the culture until the 1970s. So Wanda just wore what she had brought: simple business suits for formal interviews and long-sleeve shirts and slacks for field visits. To most Saudis, she was an oddity, but not a threat.[36]

Some Aramco officials, however, regarded her with suspicion. There were no female professionals, aside from teachers, nurses, or librarians, working for Aramco, and those few women lived almost exclusively within the company's Dhahran compound and had little, if any, contact with Saudis. "The oil industry was a very macho industry," said Marr, recalling her experience in the 1960s as the first woman on Aramco's professional staff. "It's hard today to think back on how pervasive the discrimination was back then for women," she added. "Lots of oilmen arrived there in their big boots and ten gallon hats just full of themselves. One refused to meet with me just because I was a woman."[37] Some Aramco officials were "apprehensive about Wanda," conceded Jack Butler, the Aramco official responsible for her during that visit, but she won them over with her "disarming" manner.[38]

WALKING THE HAREM GAUNTLET

Wanda admitted that she did not know what to say or do in the presence of King Saud. Since women were forbidden by Saudi law to enter the king's palace, the king received Wanda in his harem—the women's quarters—where no man except the king could go. To reach him, she had to pass through a gauntlet of several hundred long-robed, black-veiled women, all relatives of the king. They whispered and giggled as she gingerly walked on Persian carpets at least three hundred feet toward Saud. The sight of an uncovered woman, with curly hair and a Western business suit, was both fascinating and humorous. "It was my short hair that seemed to astound them most," Wanda later said. "Don't forget—they'd never seen a Western woman."

Seated on a raised dais at the opposite end of the audience chamber, the king looked formidable from a distance. "I couldn't have been more nervous," Wanda recalled. Suddenly, her female interpreter broke away, rushed up toward the king, prostrated herself in front of him, and kissed his feet. Astonished, Wanda did not know what to do. She edged forward but then stopped about twenty feet from him. Was she expected to kiss his feet, too?

Sensing her confusion, King Saud extended his hand, so she walked forward and shook it. "You are welcome," he told her. "All Americans are welcome, like my own children." She replied in Arabic but made a mistake; what she said meant, "You are welcome to my house." "Imagine *me* welcoming the king to his own house!" she later exclaimed. But it helped: "That got a laugh and eased the tension." Saud's children then came in and swarmed all over him as they each got a kiss. "It was a big family scene," she recalled, and a typical one. The king, she learned, spent that hour in the harem every day, mainly to see his children.[39]

Despite her awkward start and the press of children, Wanda was determined to get comments from the king. She asked several questions, including sensitive political queries about the Suez Canal, but all of his answers came back the same: "We are all Arab brothers and Arab brothers love each other."[40] In the end, all he gave her were gifts: a watch with a royal insignia and two harem dresses woven from solid gold thread, much too heavy for her to wear.[41]

Although Saud did not tell her anything of significance, her experience was so unusual that she gave more details about the harem visit when she returned to the United States in the spring of 1957. During a cross-country speaking tour, she was featured on local radio and television programs and in several newspapers. This "pert, pretty young American," reported the Los Angeles *Mirror-News*, was "one of the few Westerners ever admitted to an Arabian harem." At the Los Angeles press briefing, Wanda explained Saudi customs and dispelled some myths. The king had as many as four wives at any one time, as permitted by the Koran, she explained, for reasons that had "nothing to do with licentiousness." Because Saudi Arabia was a recent amalgamation of many tribes, "what's the best way to cement bonds between the various tribes and the King?" she asked. "Arrange a marriage between the chief daughter of the tribe and the ruler." When the king fathered a son by this queen, she continued, "he has given her tribe a prince of royal lineage. Whether she remains permanently in the harem or returns divorced, to the tribe, she is always a person of consequence and much respected."[42]

The king did not eat with women, Wanda explained, so she dined with one of his wives and other leading ladies for nearly two hours. They started with tea, steeped with rose petals from the king's garden, and ended with American coffee, brewed in her honor. It was a "perfectly gay hen party," Wanda wrote to her colleagues. One queen told Wanda that she was "too thin and should be fatter if I am to find a husband and have a prince." She did not mention to the queen, who thought her single, that she was actually still married. Nor did she mention the eunuchs guarding the king's harem, who "looked right through her."[43]

Although Wanda was the first female reporter to interview King Saud, his father, Ibn Saud, had met the well-known American columnist Dorothy Thompson in 1952. Wanda noted in one of her commentaries that "for the first time in history, the government honored visiting American women journalists — first Dorothy Thompson and then me — with official banquets. Mine was attended by the Minister from Syria, the Minister from Jordan, and about 40 Government officials."

The king, who ruled a country with "one foot in the 20th century, one in the 7th," she wrote in another commentary, was America's "best friend in the Middle East." Saud "found the Americans had no colonial ambitions" in Saudi Arabia, that "their commercial interests were not followed by political pressure." Although a supporter of Nasser during the Suez crisis, Saud was an uncompromising anticommunist, she insisted. While some Westerners wondered whether Saud was a Nasser "puppet," he had taken a number of steps that showed "independent leadership and balancing restraint."[44]

Wanda's columns about Saud, who would soon make his first state visit to the United States, were well received by her *Petroleum Week* readers but not by the pro-Zionist *New York Post*. That paper had actually banned Dorothy Thompson's columns in 1947 after she raised questions about Zionist influence on American foreign policy and called for a settlement in Palestine that recognized just claims on both sides. She refused to be intimidated, she said, because she was "too old and too stubborn" to yield to "blackmail," but her reputation suffered.[45]

A decade later, Wanda found herself tarred with a similar sarcasm, but mainly over a letter Wanda wrote her colleagues about the harem meeting that *Petroleum Week* had published. Under the banner headline "King Saud, the All-Arabian Boy, Just a Homebody," the *Post* poked fun at Wanda as "the first woman ever to interview Saud," whose family would be waiting for him to return quickly from his forthcoming U.S. trip to his "little old home" in the desert "just like any happy American family." The paper then chided her for being impressed by the harem's "plain, ordinary, warm home and family atmosphere — just like our own, though admittedly on a considerably larger family scale."[46] She described the king in her letter as "relaxed, warm, and smiling, not stiff like in the official pictures." As a result, "Miss Jablonski implies" that Saud was a homebody, not a playboy. The *Post*, for its part, implied that Wanda had been seduced by Saud's charm.[47]

Wanda had, indeed, gushed about Saud in that letter to her colleagues. Wanda also gave an overly flattering portrait of Saud in her columns, although they were similar in tone to a sympathetic story in *Time* magazine describing Saud as "gentle, patient and kind, with none of the deviousness that has too often given Arab politicians a bad name." Wanda's premise was, once again, that Americans misunderstood the Middle East: she disapproved of the way Saud had sometimes been negatively portrayed "as a 20th Century version of the Arabian Nights." He had inherited the throne, yet he traveled regularly to consult with village elders. He was an absolute ruler, but he was rigidly bound by tribal traditions. When oil development started in 1938, "the Saudis were still living literally on the edge of starvation, as migrant tribes in a scrubby, water-scarce desert, or in a few fly-infested mud towns." Saud, she conceded, spent "a lot of money on the court," but "nothing is more dangerous than trying to apply our Western standards to a society as totally different from ours as Saudi Arabia. This is a medieval Islamic culture, where the King is looked up to as the father of his people, and has to act correspondingly." Even on a "slow day" at court, more than a thousand mouths got dinner.[48] Though she was well known to be skeptical of oil company pronouncements, Wanda fell for Aramco's skillful mythmaking about Saud's position as "autocratic in theory but democratic in principle."[49]

BEDOUIN ETIQUETTE

Wanda's concern for Western misunderstandings of the culture and politics of the Middle East became the subtext for a number of "Petroleum Comments" that were more akin to feature articles than to commentaries. These local-color pieces—more textured than accounts that she had previously written—brought to life the setting, the history, and also the nuts and bolts of oil exploration in a very alien environment. "They were a revelation to many U.S. oil men who, at that time," Georgia Macris said, "scarcely knew where the places were."[50]

Because of oil, Saudi Arabia was changing at a frenetic pace. Walking through the mud-walled fortress city of Hofuf, "you almost feel as if you are walking in the pages of the Bible," Wanda wrote in one piece. "Then, nearby, a red sign with white squiggly Arabic writing exhorts you to 'drink Coca-Cola ice cold.'" She was amazed to see so much modernization since 1954 — drugstores "now have ear-plugs and Alka Seltzer" (both of which she used). Yet the dirt streets, ubiquitous camels, and "Moses-like" patriarchs kept reminding her that "merely 20 years ago, no more than a few dozen Europeans had ever penetrated this vast land, nearly one-third as big as the United States." Whereas Iraq, Iran, and Egypt had "felt the sweep of many civilizations," the world had moved past the barren Arabian Peninsula. Most Muslim countries had replaced ancient religious law with civil law, but not Saudi Arabia. "In some ways, life here is reminiscent of New England under the Puritans," she wrote. Sharia law, based on the Koran, banned drinking alcohol, gambling, and charging interest. Women were completely veiled in thick cloth, their face coverings ranging from a "semi-transparent black cloth for the more sophisticated women in the cities to thick cloth (with eye slits) in the towns and villages, and black masks among the Bedouin tribes." Other practices, too, including medical treatment, remained tradition-bound for many Saudis. Two fresh burn scars on the forehead of a man she met indicated his severe headache "had been 'treated' in the age-old desert fashion: applying a red-hot poker to the painful area."[51]

She particularly appreciated Arab hospitality, "a time-honored code of the desert, born of the harshness and rigor of desert life." As she traveled through the country to visit oil installations, she liked "to stop at a cluster of low-flung, black, goat-hair tents of the Bedouin." Coffee etiquette was strict here: "As a stranger, you must drink exactly three cups — less than that indicates distaste for your host's brew, while more suggests too much free-loading on your part." The Bedouin were full of grave questions: "How does one get food up into the airplane if it runs out of gasoline; why you are a journalist (the darndest question to answer); whether your children will take care of you when you grow old; is the grazing good in New York?" She tried to give serious answers, but when one sheikh asked her about the quality of grazing for sheep in her city, she simply quipped, "Adequate to our needs."[52]

She did not know what to do, however, the day a Bedouin man reacted strongly to her request to take a photograph of his wife. The man had come up to her during one of her desert expeditions, "smiling, calling out traditional, elaborate Arab greetings. 'Salaam aleikhum,' I answered. Having exhausted 90% of my Arabic, I held up my camera and pointed to the Bedouin's masked wife questioningly." No, the Bedouin angrily told her, and he suddenly dashed into his tent. For a moment she was anxious. She suddenly realized that it was culturally offensive to ask to take a photo of his wife. But "a minute later he came running back, gown flying in the air behind him, carrying a fat, baby goat." Beaming happily, he handed the goat to Wanda. She smiled and asked her guide, a drilling superintendent from California, to take photos of her holding the goat. She then tried to hand it back. "No, no, the Bedouin gesticulated firmly, pointing first to me, then the goat, then the car. Finally the light dawned—the goat was a present!"

Since the animal was obviously a prized possession, she tried to give the owner some money. To this "thin, half-starved" Bedouin, the goat "easily represented one-twentieth of his entire worldly possessions. It probably meant the only square meal for a month for his family." His "code of manners was too high," however, for him to let her leave empty-handed, and he refused payment. She tried giving coins to his child and wife, but he turned away their eager hands. Wanda was stunned. The gift of a goat, she wrote, was just as significant to her as the modern hospital she had just visited where everyone was treated for free. The desert Arabs, though scorned by more westernized Arabs as simple and illiterate, had "real character [and] are, in many ways, the true aristocrats of the Arabian world." She then added, "P.S. Anybody want a spare goat?"[53]

THE EMPTY QUARTER

Wanda also wrote colorful pieces on Aramco's ambitious drilling efforts. The latest hot spot was "right in the middle of the Rub al-Khali—the

Wanda being given a goat by a Bedouin near an oil camp, 1957.

'Empty Quarter'—a vast, Texas-size, unknown desert in southern Arabia that's still a dread name among the Bedouin." Two languages were spoken there: "Saudi Arab and Texan." With typewriter and camera in tow, she visited a group of geologists and surveyors camped in the most remote region of the Empty Quarter, "a forbidding belt of high sand mountains" in the southeast that was marked "inaccessible" on Aramco maps. "I'm told that I'm the first journalist to have flown all the way across the Rub al-Khali, and the first to have entered the sand mountain area," she wrote in the ninth column of her series.

It was an eerie place, where "the sands 'sing,' the stones 'walk,' and there are great lost cities somewhere around us. So at least say the legends." Wanda could verify only the singing. At first, she thought she was hearing the drone of airplanes. Then she heard a roar, a thunderous boom that turned into a "rumbling noise, of astonishing depth, that echoes and reechoes" for several minutes—it was a sand slide in a great amphitheater of reddish-brown peaks. The barren waste, the wild sand, howled in song.

Aramco men had recently discovered how to negotiate these treacherous skyscrapers of soft, shifting sand without camels: they outfitted one-ton, four-wheel-drive Dodge Power Wagons with DC-3 airplane tires. The young Texan surveyors careened up and down eight-hundred-foot-high slopes like "teenagers in hot-rods," Wanda noted. They offered to take her for a ride up a hillside, "a little 300-ft. job—merely the height of Chicago's 25-story Conrad Hilton Hotel." Wearing a striped cotton shirt and slim pants, she eagerly stepped up into the truck cabin, but then got nervous when her driver strapped her in with a seatbelt. "The ride up the rolling leeward side wasn't too bad—rather like an out-size Coney Island roller coaster," she wrote in her column. "Things were fine till we hit the very top—the knife edge. I got one glimpse: nothing but space. And a sheer drop of 300 feet. Somehow we slid down. How, I don't know. I never opened my eyes—and never prayed harder." They had to go fast, she later learned, because otherwise they would get stuck on the top—literally—"suspended in the middle on top, with the front wheels dangling over one side and the back wheels over the other." If they went too fast, "you're apt to go right over the knife—into nothing."

Wanda's own snapshot of the Rub al-Khali (Empty Quarter), southern Saudi Arabia, published March 8, 1957.

But not one truck had flipped over, the driver reassured her, "since the day before yesterday."

That night, the leader of the field party's Saudi army escort, Jamaal bin Umbarrik, from the warlike Murra Bedouin tribe, invited Wanda and a few of her American hosts to his traditional striped tent on the edge of the makeshift desert campground. He was a handsome man, she thought, with the delicate, elongated features of desert nomads. He wore a white flowing garment and a red-and-white checkered keffiyeh. Under the pale light of a single kerosene lantern, the group sat cross-legged on Persian carpets "sipping endless rounds of bitter, cardamom-flavored coffee in small china cups, and mint tea in little glasses encased in metal holders," Wanda told her readers. After an hour, "a gun-toting soldier-servant brought a large brass goblet, half filled with sand, topped with glowing embers. Our host reached into a pocket deep in the folds of his gown, pulled out several pieces of frankincense and dropped them ceremoniously on the embers." The servant carried the goblet to each guest. "The scent wafted into our faces. Twice it came around—

the signal to depart." She shook bin Umbarrik's hand, American-style, stooped to leave the tent, and walked slowly back across the soft sand to Aramco's air-conditioned metal trailers, which glinted in the moonlight. The juxtaposition was jarring. With a few strides, she seemed to traverse centuries.[54]

The Arab desert captivated Wanda as it had other Western travelers, including a handful of upper-class British women who had lived there decades before. Gertrude Bell, the diplomat, travel writer, and Arabist, wrote about the "irresistible chivalry of the desert." But Wanda's favorite was Freya Stark, who like Wanda had grown up mostly abroad and did not fit easily into her own society. Wanda avidly read and shared with friends *The Freya Stark Story*, an account of Stark's travels in the 1930s and 1940s. What drew Stark to the Arabian desert was "the *intimacy* with a world so strange and remote: it amounts almost to an annihilation of time." Desert life was a living experience of the Dark Ages, she wrote: "Its difference from our world went . . . beyond its permanent insecurity, which is now no longer a stranger in Europe: it was perhaps the *acceptance* of insecurity as the foundation of life."[55] Wanda, too, was drawn to this mystique—the singing sands and Bedouin frankincense.

INTERVIEWS ON THE PIRATE COAST

Wanda's next destination proved even more daunting. She ventured into the remote fringe of the Arabian peninsula along the western side of the Persian Gulf: the sheikhdoms of Qatar and Bahrain as well as the Trucial Coast states of Abu Dhabi, Dubai, and Sharjah. This was no mean feat. The hostility between these sheikhdoms and Saudi Arabia, particularly over boundary issues, was so fierce that visitors with visas for one place normally could not get them for the other. Wanda, as usual, insisted on being an exception.

At first, the roadblocks were considerable. She was, after all, a journalist and a woman. The IPC, the majors' joint venture in Iraq, never sponsored reporters' visits, one of their executives told her father's friend Brandon Grove, then Mobil's Middle East expert in London, "in view of the propensity of nearly all journalists to write things which are not approved by the Governments of the countries in which we operate."

The IPC executive would do nothing to help Wanda secure visas, but, as a courtesy to Grove, would arrange for her to visit IPC operations in Iraq and Qatar if she managed to get to them. However, the Trucial Coast and Oman, where IPC also controlled the main concessions, were off-limits because "facilities for lady visitors just do not exist."[56]

So Wanda appealed for help from another friend, Billy Fraser, a senior British Petroleum executive and son of BP's fearsome chairman, and he readily agreed. Wanda would accept, he told skeptical colleagues, "whatever accommodations can be provided." Thanks to his diplomatic skills and his powerful name, he eventually prevailed, but the process was lengthy. Since Jews were forbidden entry to many Arab states, Wanda even had to present a notarized copy of her Catholic baptismal certificate, plus authentication of the English translation.[57]

McGraw-Hill, in a press release, said that Wanda's trip was unique because she would get exclusive interviews with Arab rulers while visiting "remote places where no white woman has ever set foot." In reality, a few geologists' wives had traveled to most of the Persian Gulf emirate states, but female reporters had not, and few male reporters had, either. Many years later, IPC's senior field manager in the Persian Gulf, Paul Ensor, vividly recalled Wanda's 1957 visit to Qatar and Abu Dhabi: "I just don't remember any other reporters going around like Wanda did in the 1950s. None. No Western journalist went to the sheikhdoms. There was no outside news at all." He was "surprised to have a lady like Wanda," but she was willing to dress appropriately, with long sleeves, long skirts, and sometimes a headscarf. The rulers, for their part, were willing to meet with her, even though she was a woman and had just visited Saudi Arabia. "At that time the power of the British was such that if you were introduced by a British representative, you were OK. The sheikhs were polite with her," Ensor remembered. "I doubt very much that they had ever been interviewed before. They didn't really know what a reporter was."[58]

Although Britain had tried to control this region by establishing truces among the chronically warring states in the nineteenth century, tensions still prevailed. British oil companies—not to mention their American and French competitors—had maintained considerable interest in the area since 1932, when Standard Oil of California, known as

Socal, discovered high-quality crude on the island of Bahrain. By 1957, however, oil drilling had been far less successful in these smaller sheikhdoms (with the exception of Qatar) than in Saudi Arabia or Kuwait. Bahrain's production rate was declining, and drillers had yet to strike pay dirt in Abu Dhabi. Except in Bahrain and Qatar, the most profitable professions along the Arabian coastline were still smuggling and piracy. In native *dhows* (sailing sloops), these middlemen plied their trade in gold, cloth, European goods, and gasoline with black mar-keteers from Iran, Pakistan, and India. In Dubai, "we would sometimes see smugglers unloading gold from the Bank of England right there at the airport," Ensor said. Thus the Trucial Coast, named for Britain's hard-won truces, was still better known as the Pirate Coast.

Ensor did not know what to do with this female visitor, whose skill at extracting information was already legendary in the industry. "We had to be terribly careful with Wanda," he remembered, because of how the different shareholder companies might react to her reports. Neverthe-less, he helped her, particularly off the record, because of her "reputa-tion for not publishing what she was told in confidence."[59]

As she got to know him, Ensor proved to be an engaging English-man with considerable insight. He had worked in Iraq and the Persian Gulf since 1931 and spoke fluent Arabic. Ensor was curious to hear what Wanda had learned in Saudi Arabia because bitter rivalries between the Gulf rulers and the Saudi monarchy prevented him from traveling there (except once for an Aramco–IPC tennis tournament). He sometimes accompanied Wanda or provided a staff interpreter for her interviews, since most of the rulers had none. Between expeditions, she stayed with Ensor and his wife in Doha, Qatar's largest town, where their home pro-vided a refuge. Nonstop travel, unfamiliar food, and stressful work had exhausted her, "so we tried to leave her alone so she could write," Ensor recalled. "We had a nice room with a servant and she could get a cold bottle of beer." They even had running water—a novelty for Doha. Un-til shortly before Wanda's visit, water had been delivered in goatskin sacks slung over the backs of donkeys.[60]

Other Western oilmen were more guarded. In Kuwait, Wanda's ar-rival caused "outright consternation," Margaret Clarke recalled. A Lon-don-based public relations official for Kuwait Oil who later worked for

Wanda, Clarke was in Kuwait City for a six-week stint. She had first met Wanda in London two months earlier for a business lunch, a meal made memorable by Wanda's petulance. At a fashionable restaurant on Park Lane, "Wanda didn't like anything on the menu." She had just awakened, and all she wanted, to Clarke's dismay, was American-style coffee and plenty of it—not coffee served in a demitasse.

Wanda behaved better in Kuwait. "She cottoned on to me" at the Kuwait Oil guesthouse, Clarke recalled, because she "just wanted to spend time with another woman." A familiar female, after weeks on the road, was a relief. Kuwait in the 1950s had just one hotel, "a very seedy place, only for men," Clarke said, so the guesthouse was the only place Wanda could get lodgings. "She would keep me up until the wee hours of the morning, drinking and telling me stories about the harems she had visited." Clarke's dry wit and gregarious nature appealed to Wanda, but Kuwait Oil's managing director and his staff were "absolutely terrified" at what Wanda might learn during her visit.[61] Aside from her friendship with Clarke, Wanda found Kuwait to be less congenial than Qatar, to which she soon returned.

Qatar's oil revenue, though not as impressive as Kuwait's, was changing life in the sheikhdom. Economic conditions had been tough before the mid-1950s; the advent of cultured and synthetic pearls had undermined Qatar's former source of foreign income, pearl fishing. The country was so poor, Ensor said, "that a man would come into the marketplace with a halter around a sheep and wait until enough people gathered before he'd cut the sheep's neck and sell it off in bits." Qatar's sheikh, who received payment from the IPC every six months in the form of a truckload of silver coins, "used to bury them in a hole inside his quarters." The tobacco he smoked "was so primitive that it would send out sparks." But by 1957, increased oil production had brought not only piped-in water to a few homes, but significant cultural changes as well. The ruler, for instance, decided to pay an allowance to fathers whose sons attended school. Girls, of course, did not go to school. "That's against his principles," Wanda pointedly noted.

Aside from its modest onshore oil production, Qatar had few assets, Wanda explained in her feature on the sheikhdom (ironic, indeed, because the world's largest natural gas field would be found there in 1971).

A barren expanse of rock and sand, the Qatar peninsula "gives the impression that when God was creating the earth, he got bored with the job just about here," she quipped. "After only a few days here, you thirst for the sight of something green—even the palms of Bahrain—and for something to break the monotonous flatness—even the high sand dunes or black basalt mountains of Arabia."[62]

Ensor arranged for Wanda to hitch rides on IPC-chartered, eight-seat planes that flew regularly from Qatar to other sheikhdoms with staff, equipment, and food. The planes were old British bombers, piloted mostly by weather-beaten Australians. The aircraft were so versatile and sturdy that they could taxi up to ten miles on hard sand, but they did not fly in high winds, which delayed her visit to Dukhan, Qatar's only oil-producing field at that time. Once, on her way from Abu Dhabi to Dubai, her plane got stuck in soft sand, but she eventually got where she needed to be—just in time for cocktails.[63]

All in all, her visit to the Persian Gulf, perforce a day-to-day affair, took much longer than planned. What had been dismissed as impossible by many oil executives turned into a monthlong series of unprecedented interviews in February 1957. After a few days at her next destination, Abu Dhabi, Wanda took a sanguine view. In a four-page article, she wrote that Abu Dhabi, the largest of the seven coastal sheikhdoms, was also a desolate land, nothing but mud flats, dunes, and desert. It could not participate as easily as other sheikhdoms in contraband trade because it was farther from the opposite Gulf coast. Until the British had intervened four years earlier, "Abu Dhabi was so lawless that you didn't dare to go out of your house at night or even to travel in broad daylight without an armed escort." But "when one sees the abysmal poverty of this coast, it is easy to see what lured these people to robbery on the seas," she explained. "There is practically no fresh water; what there is is very brackish. Agriculture is nonexistent. The only steady food is fish." During stormy seasons, people survived on very little.

The few foreigners who lived there adapted. Tim Hillyard, BP's Trucial Coast representative and the only Westerner with a wife and child in Abu Dhabi, served Wanda a local delicacy—"sea cow," a rare sea mammal that tasted like veal. "Hillyard claims the natives capture the sea cow by jumping on its back and stuffing mud into its nostrils,"

Wanda wrote, "and says the creature actually 'moos' like a cow and eats grass on the seabed." Striking a major oil field, she concluded, was Abu Dhabi's best hope. (Only one year later, the majors discovered a massive offshore oil field, Umm Shaif.)

The Hillyard family took Wanda to her first meeting with Abu Dhabi's ruler, Sheikh Shakbut ibn Sultan. Wearing traditional robes and an impressive sword at his waistband, Shakbut awaited them in the courtyard of his simple, mud-walled palace. "In a semicircle stood several of his brothers, plus a dozen or so lesser sheikhs and bodyguards—armed with rifles, pistols, and daggers," Wanda wrote. This sheikh had ruled for twenty-eight years, a remarkable feat given that ten of the previous twelve rulers of Abu Dhabi had been assassinated. Still, she was taken aback by the array of weaponry. Suddenly, the Hillyards' three-year-old daughter, Deborah, walked up to Shakbut by herself, reached up to shake his hand, and "piped, in a high but firm baby voice: 'Salaam aleikum.' The ruler's impassive face creased with delight" as he leaned down to take her hand. Deborah then proceeded to greet each man in the semicircle the same way. After that performance, Wanda relaxed.[64]

Within minutes of meeting her, Wanda reported, Shakbut asked her, "in rapid succession, whether Venezuelan oil was higher grade than Persian Gulf oil, how much U.S. oil was being shipped to Europe this month as against last month, why all the American drillers he had met were from California, and how many barrels of oil you would get if all the coal of the world was converted to oil?"

Shakbut was also curious about the United States. How big was her country? About three times the size of Saudi Arabia, she replied. Then how do you have enough room for all those factories, and how do you keep those millions of people from "bumping into one another?" As in many of her columns, Wanda used dialogue to illustrate how savvy Middle Easterners were—and how ignorant of Western culture. She then added a warning: "For all his personal charm, Shakbut is not an easy man to do business with. Any oil men or British political officials who think they can go in and tell him what to do quickly learn otherwise." He was an astute negotiator who almost never compromised.

In a crisis, though, the sheikh was ready to help. A few months before, when the British invaded Suez, riots forced the banks in Bahrain to

Wanda with Sheikh Shakbut ibn Sultan, ruler of Abu Dhabi, in the courtyard of his palace, published March 15, 1957.

close. Hillyard, anxious to pay his local workers on schedule and keep up a semblance of normality, asked Shakbut if he would lend him some money—£15,000 ($42,000). "A few hours later, the money was delivered to Hillyard, all in cash, wrapped in an old rag," Wanda wrote. "No receipt was asked." Like other sheikhs, Shakbut kept much of his wealth—mostly gold and silver coins—stashed under his bed or buried in the ground of his mud palace.[65]

In one of the few television interviews she gave in the 1980s, Wanda let on that Shakbut also had a sense of humor. When she dared to ask him, "Is it true that Abu Dhabi is the center of slave traffic?" he looked her up and down and replied, "You needn't worry. You're too old and too skinny." Later, he grew more openly provocative. Looking over at BP's Hillyard, he said, "The British have drilled for oil here and have found nothing." Then he spat on the ground. "I know the Americans. They have found oil for Saud." He spat again. "And the Americans, they found oil for the Khalifas of Bahrain." And he spat again. "And I know it was the Americans who found oil in Kuwait for the Sabahs." And he spat once more. "So don't you think I should throw out the British and bring in the Americans, eh?" That anecdote, with Shakbut skewering her host, she decided not to print.[66]

She also did not recount what was for her an embarrassing episode, one that Ensor later recalled with amusement. When Wanda interviewed the sheikh of Bahrain, she knew that local coffee etiquette was unique, but she forgot a key signal. In Bahrain, "you are served coffee in a little cup," Ensor explained. "When you have had enough, you shake the cup. Poor Wanda could not remember the etiquette. She remembered you had to shake something. She shook her head, her hand, but nothing worked so she ended up drinking fifteen cups or more of this very, very strong coffee. They served only a tiny amount but it was laced with cardamom and other things. It sat in a pot on embers of charcoal all day."[67] Other customs were a challenge as well. When Wanda had lunch with Shakbut's wife, Miriam, they met in her bedroom—"a simple, white-washed room, with a high-legged western bed." Sitting on Persian carpets, they ate the traditional fare, "boiled sheep on a high mound of tasty, saffron-colored rice, mixed with nuts and raisins," set out on a huge copper tray. "We ate in the customary fashion: digging in with our right hands (using one's left hand is very bad manners, indeed)—I, in my ineptitude, dribbling rice all up my arm, which Miriam politely ignored."[68]

Around her stories and character sketches, Wanda wove the dry data she knew her readers needed: Aramco's drilling sites, the latest test results, the depth of the "pay zone," the rate of production, the quality of the crude, the water-injection plans. By 1957, Persian Gulf countries alone were supplying almost one quarter of the world's oil supply, up from 17 percent in 1951. Wanda reported, for instance, that a massive expansion program in the Neutral Zone, the "small, featureless strip of scorching desert" sandwiched between Kuwait and Saudi Arabia, was so successful that production would quintuple that year. She often used American terms of reference: Kuwait, "this little Connecticut-sized" state that ten years ago had just "a lone strip of asphalt and two radio pylons" as tangible signs of modernity, boasted one of the world's biggest oil fields, the world's largest loading pier, larger oil reserves than the United States, and oil revenue that rivaled Saudi Arabia's.[69]

As always, Wanda sought to debunk myths. "Many people have the impression that all you have to do in the Middle East is scratch the ground and oil comes gushing up," she noted. Kuwait might be a bonanza, but not the Pirate Coast, where intensive exploration had led to eleven "exceedingly dry holes." Exploring for oil was a high-risk, high-cost business. When she asked Paul Ensor how he felt about the lack of success in his region, he wryly retorted, "How would you feel in my shoes?" In Bahrain, the first Persian Gulf discovery turned out to be just a "U.S.-sized oil field," and despite further exploration, its proven reserves in 1957 were "mere peanuts" compared with Saudi Arabia and Kuwait. These results had strained relations with Bahrain's ruler, Sheikh Sulman, who envied his neighbors. Wanda learned this the hard way during an interview. When she "undiplomatically" asked Sulman about the Saudis, she ruefully reported, "the audience, which up to that point had been a pleasant, chatty meeting, ended abruptly."[70]

She also conveyed the excitement and weariness of Westerners working in such an alien environment. Oil was a business driven both by science and by hunches, where jinxes, mysteries, and guesses were as much a part of the work as derricks and seismographs. Camaraderie was essential to surviving the harsh working conditions, yet the labor force was often so diverse that "it takes a linguist to get along." Wanda delighted in the oilmen's humor. A major discovery in Kuwait's desert was called Raudhatain, meaning "Twin Gardens." "When there is an occasional heavy rain," she noted dryly, "two spots in the sand turn green."

The weather, of course, was a challenge. Wanda quoted one Texan who defied anyone to find him a spot hotter than the Neutral Zone: "It's not only a lot hotter than Houston, but even more muggy." The sun was so brutal in summer "that the lizards run up the hot water pipes at the oil camps to cool off." The Qatar peninsula was equally steamy, and the enticing turquoise water offered little relief. "The shallow parts are like hot soup," Wanda wrote. "The deeper, cooler water abounds in shark and barracuda." Sometimes a howling wind darkened the skies, blistered the skin, and even swept sand through tightly sealed doors and windows.

In these eighteen articles spanning the winter of 1956–57, Wanda wrote about the look and feel of the places she saw, the personalities of the rulers she interviewed, and the customs and concerns of the people she met. She used quotes, dialogue, details, and vignettes to knit her accounts together. Keenly aware of American stereotypes, she humanized the Arabs by describing each individual and giving him or her a voice. She recounted many of her stories without commentary, letting the Arabs speak for themselves. When she occasionally set out her views, she empathized with the predicaments and opportunities that oil had caused. She appreciated what the oil companies were doing under trying conditions, but she challenged their insistence on total control over production. The Arab perspective could not be simply ignored.[71]

For Westerners who appreciated Arabs, Wanda's tone was a tonic. "Thank you for giving the Ruler and his point of view a square deal in such a refreshing style," wrote Tim Hillyard. His wife, Susan, also wrote to tell Wanda what had happened when she took the packet of press clips that Wanda had sent for the harem, including photos of the women. "The girls were thrilled," Susan Hillyard wrote, but they promptly stashed the packet under a carpet "lest unauthorized eyes should spot it" or the sheikh himself take it from them.[72]

Fortuitously, many of these columns were published during King

Saud's visit to the United States. The king made headlines for both political and personal reasons: Washington was courting his support, and
Saud won sympathy for bringing his young crippled son to the United
States for surgery. The two made for appealing front-page photos. Yet
compared with Wanda's stories in *Petroleum Week*, mainstream U.S.
press coverage of Saud and Middle East oil issues was superficial.[73]

Wanda was more than generous in her characterization of Saud and
the Persian Gulf leaders. She was so taken with their culture, awed by
the starkness of their countries, and admiring of their ability to hew to
tribal customs while trying to adjust to twentieth-century pressures that
her bias showed. Captivated by the utterly alien ways of the desert, she
found that the longer she stayed, the more she wanted to focus on "the
feel of the place."[74]

THE IRAN-IRAQ VIEW

Saudi Arabia and the Persian Gulf gave her so much to write about that
Wanda did not reach Iran until March 1957, nearly four months after leaving New York. Compared with the Pirate Coast, Tehran seemed
familiar, but the mood was quite different from that of her first visit
in 1953–54. She could tell that something had changed as soon as she
walked into the Park Hotel: American, British, Dutch, and Italian oilmen were "swarming the lobbies," she reported, and rumors of imminent new exploration deals were "thicker than ants at a picnic."[75]

It did not take her long to find out what was going on. Hussein Ala,
the minister of court whom Wanda had charmed in 1954, was again
prime minister, and he wanted to see her right away. Key decision-
makers at Iran's national oil company, the NIOC, also sought her out,
eager to tell her about their first discovery in the Qum region. Emboldened by this find and a surge in Abadan's refinery output, they knew just
what to do with the swarming foreigners: ask for better terms on any new
exploration deals—better than what they had conceded to the consortium in 1954. In Wanda, they knew they had a sympathetic listener who
would report their views to the West.[76]

But the rumors of imminent deals were exaggerated, she cabled immediately to New York. They belied a fundamental conflict: the NIOC

wanted 50 percent participation in new exploration projects, plus fifty-fifty profit-sharing. To foreign oil companies, this meant Iran would reap 75 percent of any profit—an outright violation, in their view, of the fifty-fifty split that prevailed in the Middle East and in Iran's own contract with the consortium. Operators would have to take huge risks in unknown areas, so the majors were, at least so far, "indicating thumbs down" on any NIOC partnership. The Italians, however, had already broken ranks by initialing a deal, and several medium-sized American companies were still negotiating.[77]

Wanda gathered the information and staked out her position in a column. The Iranian thinking "has a certain philosophical consistency," she wrote. It was logical: the NIOC wanted to be treated as a viable oil company, working directly as a partner with foreign companies and taking the same risks and responsibilities. The NIOC had earned that right through the expertise of its personnel and its track record. Whereas the consortium settlement was the best deal that Iran could get in 1954, the tight oil market in 1957, in the wake of the Suez crisis, strengthened Iran's bargaining stance. Although the NIOC was willing to finesse the deals if there were tax problems in a company's home country, direct participation was a sine qua non. "Americans shouldn't judge everything according to the U.S. way of life, where the government is the servant of the people," one Iranian official told her. "Here the government is much closer to being all-powerful." Having the NIOC as a partner would be an advantage to any foreign oil firm. The Iranian government ultimately wanted to make the NIOC an international oil company, but it was ready to start with a partnership. Once again, as she had in Venezuela in 1948 and Iran in 1954, Wanda articulated a compelling argument that reflected Perez Alfonso's position: oil-producing countries were justified in seeking greater participation in the development of their own natural resources.[78]

At her next stop, Baghdad, Wanda also found government leaders eager to talk with her. Interviews with two leading Iraqi cabinet officials gave her plenty of quotes on how to ensure the security of Middle East pipelines. Although Syria had sabotaged IPC's lines during the Suez crisis, the Iraqis wanted to negotiate treaties with neighboring states rather than rely on Western powers to keep order. In her column,

Wanda outlined the oil companies' coolness to any pacts between Middle East countries. An agreement just between a company and the government was no longer enough, Iraq's economics minister, Nadim Pachachi, told her. "It is not just a private matter involving a private oil company's profits. It is a matter that affects the welfare of the whole state."[79]

Wanda managed to get an Egyptian visa, so she went to Cairo, hoping to interview Gamal Abdel Nasser. But unable to get an immediate audience, she flew home to New York. Her office had been pressing her to come back to attend an April 17 press awards dinner in Washington, sponsored by the Associated Business Publications and judged by Northwestern University's Medill School of Journalism. Wanda won the Jesse H. Neal prize, known as the Pulitzer Prize of the business press, for editorial writing in 1956.[80]

LEARN HOW THE ARABS THINK

Upon her return later that April, Wanda felt obliged to take on what for her was a most difficult assignment—much harder than the Arabian sand mountains or the Pirate Coast. McGraw-Hill wanted her to go on a national speaking tour that included numerous television and radio interviews. Quick-witted and tart-tongued, Wanda was thought to be a naturally gifted speaker. She was certainly a good storyteller, but talking in front of a large audience terrified her, so much so that her editor had paid for her to take a Dale Carnegie course in public speaking the year before. Though exhausted from illness and travel, Wanda knew she had to go.[81]

But before she left, she had to come to terms with something even worse: her failed marriage. In November 1956, just prior to her departure for the Middle East, she learned that her husband was having an affair with a secretary at his law firm. Wanda had known for some time that she had alienated Jaqua, her friend Eleanor Schwartz recalled, but the end was still a shock. "She came close to having a breakdown" when she realized the marriage was over. "She was very lonesome, desperately in need of company. She'd call me up in the middle of the night to talk." She used to take long walks in the rain trying to decide what to do, she told another friend. Aramco's Jack Butler also remembered how devastated she was. When she arrived in Dhahran in December, he came to pick her up at the guest house and found her crying. "She was very upset. I really saw her vulnerability then and I never saw her like that again." From Jaqua's standpoint, the two of them "had pretty much parted company" before her Middle East trip.[82]

When she returned home in April 1957, Jaqua insisted on a quick divorce. His girlfriend was pregnant. So after she finished her spring speaking tour, Wanda and Jaqua spent six weeks in Reno, Nevada, to establish residency and finalized the divorce on July 31, 1957. Jaqua immediately married the secretary, who soon gave birth. Eleanor Schwartz recalled that Wanda "went through a bad time," but Georgia Macris said that Wanda buried her emotions and did not show her wounds at work.[83]

Indeed, the speaking tour, Macris recalled, was a real success. "Oil executives and wildcatters alike fell totally under her spell." A Texan friend and sometime critic wrote to say that Wanda did a "top-notch job" on an NBC interview show, adding "it's perfectly clear now that you were pulling my leg when you once told me that you were afflicted on occasion with stage fright." Her "off-the-cuff approach" worked, a colleague also noted: "To say that she 'knocked them dead' at the Security Analysts meeting is to put it mildly."[84]

In every venue, Wanda spoke about the threat of a Soviet takeover in the Middle East and about America's lack of understanding of the Arab perspective. At the World Affairs Council meeting in San Francisco, when the chairman introduced her as keynote speaker and asked for the title of her address, she quipped, "How to Develop Communists and Alienate Arabs." She took nearly an hour to state her case.

Most Americans, she explained, did not realize that many Arabs thought of themselves as part of a fairly homogenous group, rather than as citizens of separate countries, so Nasser's call for Arab unity had real appeal. Only ten years earlier, the United States was well respected by Arabs, but its support for the creation of Israel had changed that. Recent insulting incidents, such as New York Mayor Robert Wagner Jr.'s discourtesy to King Saud and ill-timed support for controversial Israeli actions, contributed to growing disillusionment among Arabs. Radio

Moscow was taking advantage of these Western "gaffes," and its propaganda was heard all over the Middle East thanks to the proliferation of radios. As a result, Arabs "do not think the United States wants to save the Middle East *from* international communism, but wants to save it *for* the United States." What's more, some leaders were reading Western bank forecasts of higher oil demand, which strengthened their bargaining position. Given the region's importance in keeping communism at bay and fueling Western industries, the United States had to rethink its policies. An important first step would be to "learn to think how the Arabs think, not how *we* think they should think."[85]

That lesson had been a major focus of Wanda's reporting from 1954 onward. That she was an expert on the oil industry, not a Middle East scholar, did not stop her. Arab nationalism would evolve, but Wanda challenged the oil industry to think beyond its ethnocentric parameters and give more consideration to the cultural implications of its business decisions. Although touched by the romance of the desert, she brought a dose of reality to the Western oil world.

5.

The Advent of
Oil Nationalism

On December 22, 1956, Ahmed Zaki Yamani, a young Saudi lawyer with two American law degrees, went to Abdullah Tariki's home in Jeddah to consult with him on a contract. He was surprised to find his boss talking animatedly with a Western woman. "I didn't know who she was," Yamani recalled. "I was introduced to her and we shook hands. That was all. But as I was leaving, someone took me aside and said, 'Do you know who that woman is? She's a spy for the oil companies. She's trying to get what is in the mind of Tariki.'" Tariki, however, was unperturbed. Though quite suspicious of Westerners, he later told Yamani that Wanda was worth the risk—a gold mine of information. "He didn't think Wanda was an agent. 'She is pro-oil companies, but we can pass messages through her to them. She can help us.'"[1]

That it was the Christmas season seemed incongruous to Wanda in the desert setting. "The temperature is 88 degrees, the Red Sea is blue, it's awfully humid, and the Santa Claus at the door of the American Embassy compound looks wilted," she wrote her editor. Tariki generated even more heat: "He's quite a ball of fire—am sending you a column on him." That column, which appeared in February 1957, was the best known of the eighteen she wrote during her five-month 1956–57 trip, and it was the first full-length feature on one of OPEC's two founding fathers. Wanda gave Tariki "exposure and credibility in the outside world, in the West," said Yamani. "She was important for both Tariki and Perez Alfonso. The companies didn't pay attention to their views back then."[2]

But the significance of these reports went well beyond bringing Tariki to the fore. Wanda's frequent trips to the Middle East and Venezuela in those years produced the most extensive coverage in the Western press on the emergence of nationalism as a serious challenge to the seemingly unassailable oil club. In the process, she educated all sides—oil nationalists as well as Western executives and diplomats—about the complexities of the issues at stake. She also expanded the range of influence that a business reporter could exert, both in print and in private.

SON OF THE DESERT

Wanda quickly sized up Tariki in that controversial first column: "This well-read, proud, engineer-geologist son of a Bedouin of the desert—the 'new generation'—may not be widely known outside of Saudi Arabia. But his policies may have far-reaching repercussions." During three days of talks at his office and in his home, Tariki told Wanda why he wanted Aramco to develop a better partnership with his country—and how.

Both the "why" and the "how" proved quite a shock. Tariki's first goal was to get a genuine fifty-fifty profit split by insisting that Aramco stop underpaying. He accused the company of distorting its obligations through improper accounting and exemptions. He then outlined a more fundamental change: Aramco should be transformed from an exploration and production subsidiary into a full-fledged oil company by expanding into refining and marketing in Saudi Arabia, and then adding tankers, foreign marketing, and even service stations abroad. When Wanda told him that U.S. antitrust laws prohibited that kind of collaboration between the four owners, Tariki replied, "Why should we suffer because of U.S. laws? They don't apply in Saudi Arabia." Tariki also resented the majors' lack of openness on pricing issues. "Do you understand my point?" Tariki asked her, leaning forward earnestly.

"Partners are supposed to share." It was not right for Aramco's parent companies to determine Saudi oil prices and policies based solely on their needs, not Saudi Arabia's. Although Aramco had recently moved its chief officers to Dhahran, the "real show" was still "run out of New York" by an executive committee in the interest of the companies, not the people of Saudi Arabia. Tariki then gave Wanda "a significant clue" to his more radical long-term plans: If Aramco could not, under U.S. law, become a fully integrated company, "maybe the owners should split up the concession." For the Saudis, it might be better to have four competing companies rather than four owning one.

Wanda also gave readers a feel for Tariki's personality. He "combined bluntness with great personal charm," she wrote. "He can say the most drastic things with the most pleasant smile." Remarkably candid for a government official, he admitted, for instance, "that he hasn't got very far" in trying to develop joint oil policies with Kuwait and Iran. His background gave clues to his character: the son of Bedouins from the Najd desert, he was proud of his education in Cairo and his master's degree in petroleum engineering and geology from the University of Texas. He had worked as a Texaco trainee in Texas and California, he told Wanda. And how did he like West Texas? "I liked the wide open spaces," he replied. How about California? "'They were more tolerant of a Mexican than the Texans'—an obvious allusion to the American habit of regarding all brown-skinned people as Mexicans," she added pointedly.

Tariki's lifestyle reflected two contrasting cultures. He lived in a Western-style villa but wore the traditional Arab robe and keffiyeh. His home showed his "softer side," she thought. "The moment you enter his walled garden, his intense love of animals and flowers is immediately apparent." Oleanders and carnations bloomed in borders. He delighted in showing off his chestnut Arab mare. At one point, Wanda was "almost run down by a baby gazelle, being chased 'round and 'round the garden by two sleek hunting dogs, with Tariki, like a young boy, laughingly joining the chase. Turkeys calmly walk around in the bedlam." She let the scene speak for itself, along with a striking photo of a smiling Tariki holding a young gazelle in his arms. The picture became well known in international oil circles, and the subtext for the cold war audience was obvious: this was no sinister communist. For Tariki, the oil discoveries were "giving the Arab his first real chance to build a better life in the desert." She then added her own coda: clearly, this son of the desert wanted to build that better life.[3]

HONING THE ARGUMENT

Later in 1957, however, Wanda voiced Aramco's perspective in a follow-up column. "Tariki—who's quick and bright, but somewhat inflexible, and inexperienced in the intricacies of the international oil business," she wrote, "has been coming up with some startling demands." His plans to exert Saudi control over Aramco were commercially untenable; Aramco's owners could not be expected to change their networks "just to please Arabia." The Saudis had no right to tax profits on the refining, distribution, and marketing of Saudi oil outside their country, just as a British textile mill did not pay income taxes to the U.S. government for fabricating cloth from American cotton.[4]

Wanda sent Tariki a pre-print copy of this column. She wrote, "I don't expect you to agree with it—any more than I expected the Aramco group to agree with my earlier column quoting your side of the story," even though he had just written to her about how much he enjoyed her articles. "However, having met you, I'm confident that you'll understand that this column is not intended as a personal disagreement—but simply as an attempt to be fair in reporting both sides of a controversy."[5]

MacPherson also got advance notice. "I don't imagine you'll agree with it," she wrote him, too, "but I thought you should see it anyway." MacPherson, however, had cut back his advisory role with the Saudis, in part because of Tariki. "I was fascinated to hear about your problems with our friend Abdullah—he sounds pretty ungrateful," she told him. "Without your help he wouldn't have been much." An unsigned *Petroleum Week* article—most likely hers—noted that MacPherson had worked closely with the "young, tough" Saudi oil director, but "just what influence he has had—if any—on Tariki is difficult to ascertain, if only for the reason that those who know Tariki best doubt that he is often open to suggestions."[6]

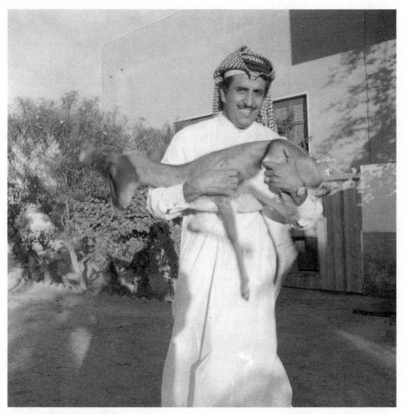

Abdullah Tariki and gazelle at his villa in Jeddah, Saudi Arabia, December 1956.

This critical view of Tariki mellowed in the spring of 1958. Although Wanda's column on Aramco's rebuttal was "rather dimly received by A. T.," Wanda learned, she wanted to see Tariki again because there was big news out of Saudi Arabia: Crown Prince Faisal and other royal family members had effectively deposed King Saud in March for, among other reasons, reportedly financing an assassination attempt on Nasser. What's more, Saud's ill health, inattention, and lavish spending had brought his country near bankruptcy. Tariki survived the upheaval with his job intact, but Wanda wanted to see for herself—something most Western journalists simply could not do, especially on short notice.[7]

Even though Tariki was annoyed with Wanda, he authorized her visa. When she reached Jeddah in late April, she wired New York that despite articles in the U.S. press about Faisal's anti-Americanism, "there are no signs of it here." In fact, she found evidence to the contrary. Faisal's gestures toward Nasser and Arab nationalism were more symbolic than substantive. Instead, he publicly supported good relations with the United States and halted arbitration proceedings against Aramco on a pipeline dispute that Tariki had initiated. "My instructions from Faisal are to be reasonable and fair," Tariki told her, "and to protect not only Saudi Arabia's interests but also Aramco's." Faisal "wants us to get along." In effect, Tariki publicly admitted in *Petroleum Week* that Faisal had told him to tone down his criticism of Aramco.[8]

That was a relief for Aramco, but two weeks later, Wanda wrote a column that the company must have considered unfriendly, even shocking. The crown prince wanted to avoid arbitration, but his oil administrator wanted to redefine his country's role in the petroleum business. So who, Wanda asked, was determining Saudi Arabia's oil policy? It was difficult to assess, she conceded, whether Tariki had the full support of Faisal and the new council of ministers, but Tariki's position was critical: "Under the present set-up, what Tariki thinks counts, for he is the man with whom final terms must in fact be negotiated."[9]

Tariki's game plan—the "key to all his thinking"—was to get foreign marketing experience so the Saudis could eventually ensure that oil exports conformed to "Arab policies." His country's future, he said, depended on oil, its only major natural resource, yet foreigners controlled it. "Our rights end at the loading docks," he complained. But that would change. He hoped to modify Aramco's concession and make sure that any new ones had better terms. If the government could get some royalties paid in oil, not dollars, Saudis could learn to market that oil and gain enough experience to run their own oil company. The Western press claimed incorrectly that Tariki would nationalize Aramco. "Nothing is further from the truth," he told Wanda. "While I can't predict what might happen ten years from now, right now we couldn't succeed in nationalization even if we tried." Saudi Arabia had to cooperate with the West "because that's where our markets are." But he hinted at an underlying threat: Aramco should get "in tune" with Saudi aspirations so his countrymen saw it acting in their interests, too. If not, "there'll be a blow up eventually and there'll be nationalization. And that will hurt us, as well as the companies."[10]

These were inflammatory words, but the argument was straightforward. This was the first time a Saudi had explicitly articulated a step-by-step process for taking control of the country's oil industry. Wanda refrained from commenting directly, but her tone was one of interest without condescension. She had, in fact, helped him frame his argument, just as she had synthesized Perez Alfonso's ideas in 1948. As she later explained, Tariki "did not express himself very well," so she helped him phrase his statements in proper English, both in 1956 and 1958. After talking for hours, "I would go off and write what I thought he had said and then show it to him the next day, and he would say yes or no." The process went on for days. For *Petroleum Week*, this was a major scoop.[11]

Feisal Mazidi, a young, British-educated Kuwaiti official, later explained the significance of Wanda's guidance: "She helped us choose the right words and phrases. Because the oilmen looked down upon us as unintelligent, her help was very important. You could trust her judgment. She would not speak to us directly about her mind but she saw that we deserved to have power over our own oil. Wanda so annoyed the oil companies once she got hold of me and the others." She told friends she was not taking sides, just trying to be fair: English was hard for the Arabs. But her discussions with Tariki also gave her greater insight into the Arab mind-set. While he refined his arguments and moderated his immediate goals, she became more sensitive to his perspective.[12]

TUTORING THE "SWARTHY" ARAB

Wanda's extensive, generally sympathetic coverage of Tariki differed markedly from the scant and mostly hostile reporting on him in the mainstream U.S. press. The *Wall Street Journal,* in its first brief mention of Tariki, in August 1958, described him as "Saudi Arabia's power-hungry, pro-Nasser director general of petroleum." *Time* first took note in October 1958 of the "swart, smiling" Arab oil expert whom American oilmen most respected and feared. Although from a land of sheikhs "with Cadillacs and concubines," *Time* said, this "Nasser admirer" was "personally incorruptible" but favored "Arab control of future oil marketing." The *New York Times,* in a rare mention, reported Tariki's "complaint" that Aramco used hidden discounts to minimize payments to Saudi Arabia. It also ran a remarkable Associated Press report in mid-1958 that described how a "smiling, swarthy" Tariki had high hopes for a constitutional monarchy—a subject that contributed eventually to his banishment from the kingdom, and one that Wanda deliberately avoided. She thought she had enough controversy just reporting on oil.[13]

Fortune, on the other hand, gave Tariki considerable attention in May 1958, eighteen months after Wanda's first full-length feature. This leading U.S. business publication described Tariki as "a fanatical nationalist" and "foolish" Nasser-follower for whom money was less important "than the power he seems to crave." The oil companies still had the upper hand, but Tariki was a threat: "Imagine a nationalist oil minister who conceives of oil as an instrument of Arabian policy as well as a source of revenue and who possesses a sharp understanding of the strategy of world oil." He might bring on a price war or try to set production limits, "a kind of General Thompson of the Middle East." (This was the Texas bureaucrat whose commission set production limits in the largest U.S. oil state and disparaged Wanda in testimony before a congressional panel for describing the U.S. market as partly regulated.) Six months later, though, *Fortune* reported that Aramco's nationalization was not imminent. Its point of reference: *Petroleum Week*'s June 1958 report from Jeddah, in which Tariki said Saudi Arabia had no interest in nationalizing Aramco because it needed Western oil markets.[14]

For Tariki, Wanda herself became a point of reference. During their meetings in 1956 and 1958, he plied her with questions about the secrets of oil pricing, the inner workings of the Aramco majors, and the policies of the U.S. government. Over the next few years, she continued to meet him for lunches and late-night talks in Beirut, Cairo, and New York. "Wanda developed a wonderful relationship with him," recalled one of Aramco's few Arabic-speaking officials from that time, Mike Ameen. "He was difficult with oilmen but talked openly and frankly with her. She was a good sounding board." She liked Tariki, yet shrewdly benefited from their friendship by gaining exceptional insight into an up-and-coming challenger.[15] Tariki also relied on two other Westerners in the 1950s, the irascible MacPherson and former Jersey lawyer Frank Hendryx. But whereas Hendryx tended to go along with whatever Tariki said, Ameen recalled, "when Wanda heard something from Tariki that was off the wall, she would tell him it was—no doubt about it." She became a "kind of mentor" who "introduced Tariki to the oil world," Ameen said. "Wanda knew a lot more about the oil business than he did."[16]

Tariki not only benefited from her tutoring but also from her network of contacts. Wanda shared her "enormous" anecdotal memory with Tariki, recalled J. E. ("Jack") Hartshorn, then the senior Middle East editor at the *Economist*. "She could always remember who had done what on behalf of whom." She would even correct Hartshorn— "actually it really happened this way," she would say, and, he added, "she was always right." She also introduced Tariki to Richard Nolte, a researcher who would write favorably about him in *Foreign Affairs* and later became U.S. ambassador to Egypt. "She mentored Tariki—and me as well," Nolte later explained. She had "a teacher complex—it was in her nature to inform and instruct people."[17]

She also understood Tariki's difficulty with fitting in wherever he lived. Not only did he have trouble in Texas, where he was thrown out of public places because he was thought to be Mexican, but he had been rejected when he returned to Saudi Arabia in 1948 with his American wife, Eleanor. As the first Saudi with a U.S. master's degree in petroleum engineering, he was sent to Dhahran as his government's chief representative to Aramco, but the American community snubbed both him and his wife. Despite Tariki's rank, Aramco initially denied him housing in the senior officers' camp, and the Tarikis were refused entry into certain Aramco facilities. The prevailing attitude, recalled one senior Aramco executive, was "We're in charge here, this is our company."[18]

What's more, his wife caused her own trouble. "Prior to their divorce in 1954," a 1970 CIA report noted, the Tarikis "survived numerous publicized incidents involving the indiscreet behavior of Mrs. Tariqi [*sic*] with American Aramco employees and some members of the royal family." Widespread rumors of her affairs were tough on Tariki, said Ameen, whose wife, Pat, occasionally took care of the Tarikis' son. "He got teed off at the way everyone was treating him. The anger just built up inside. He had a real chip on his shoulder. But Wanda knew just how to talk with him."[19] She, too, was an outsider, the stranger in every school, the "unfriendly alien" in New York, the lone woman at oil banquets or in the bars of Beirut. Given how open Wanda was with Tariki, she may well have told him of some of her own marital woes.[20]

Onnic Marashian, *Petroleum Week*'s Beirut correspondent, took mental notes as he observed Wanda and Tariki together. "Wanda and Tariki were very close," he recalled. "You could see that in the way they talked with each other, in their body language. She was devoted to Tariki, if she could be devoted to anybody. She really liked him and found him fascinating." Marashian did not think she was intentionally helping the Arab cause or deliberately promoting Tariki, but he thought she did just that by revealing so much to him about the way the companies worked. He watched her in action at the Hotel St. Georges and the Normandie, cigarette and scotch in hand, needling Tariki and oilmen alike. One of her techniques was to "make light of things to provoke an argument, to get people to talk and defend themselves" as a way to find out what was really going on. "I was fascinated by the way she kept asking questions in different ways, how she would argue with such high-level sources." Driven but not confrontational, she "debated in good spirit for the sake of the argument, not to try to 'get' people."[21]

Wanda mentored Marashian as well. From New York, she would dash off memos with news tips and journalistic advice. "Open-mindedness is still the greatest asset in this business," she wrote to him. Being sensitive to all views was critical. As Georgia Macris wrote to Marashian, "Cultivating the gardens" on both sides of the fence "equally assiduously" was "essential" for Wanda.[22]

THE MESSENGER

Only later would Marashian learn that Wanda was helping Tariki and the Aramco companies behind the scenes in a more deliberate way. Communication between the two sides was terrible. Aramco and parent company executives "basically tried to ignore Tariki and his calls for greater control," Aramco official Jack Butler said recently. "Top management would not talk with Tariki. Our channel was direct to the king" or the finance minister. Aramco officials eventually recognized the inadequacy of this policy, conceding in one confidential memo that they did not know "what channels to follow" to get business done—a memo written in 1957, nearly a year after Wanda began her extensive contacts with Tariki. She forced senior management to pay attention to him, Butler said. Aramco officials dissected and debated everything she wrote on Saudi Arabia, but her articles on Tariki caused the greatest stir.[23]

Tariki complicated relations considerably. He so "hated the oil companies," Yamani recalled, that "when he became oil minister [in 1960] ...he refused to receive letters from Aramco" or respond to them directly. All Aramco correspondence had to be sent to his assistant, Abdulhady Taher, who sent it on to Tariki. He then drafted responses and rerouted them through Taher for signature. When Yamani replaced Tariki as oil minister in 1962, Butler came to his office to request a meeting for Robert Brougham, Aramco's president. Yamani's secretary suggested he make the Americans wait a while, the way Tariki would, but Yamani refused. "That day was a turning point," Yamani recalled. "I told the Americans they were welcome anytime."[24]

But while Tariki was in office, Wanda became, in effect, a messenger. She "talked sense to" both Tariki and Aramco, Yamani said. "Tariki, who didn't have any real contact with Aramco, found Wanda to be a good intermediary." She "put light on his views" since she was already "very taken with Arab nationalism." Tariki had legitimate arguments, she told Aramco; nationalism was not just mob psychology. Wanda "was very blunt and direct and trustworthy," Butler explained. There could have been a serious confrontation, but Wanda "helped keep a constructive dialogue going—a dialogue among people who would not normally have communicated with each other."[25]

After returning to New York in June 1958 from her six-week trip to Jeddah, Beirut, and Cairo, Wanda accepted Jersey's request to meet with her, but she insisted on speaking to the entire board as she had a decade before, not just individual directors. She was blunt, as usual: Pan-Arabism had made substantial strides since her visit the previous year, she told the board. She found "almost universal adulation for Nasser among Arabs of all classes, and a hostility towards the West that had deepened markedly." There was a "growing outcry against the absentee landlordism" of the majors. According to A. Franklyn Williams, a British diplomat who debriefed her after the meeting, "She had listened to many a bitter diatribe against those international oil companies who, from foreign capitals, were draining off the wealth of the Arab countries! It was intolerable that, from their remote fastnesses in London, New York, Pittsburgh, etc., top executives of oil companies should control the economic destinies of Middle East oil producing states." Officials from Aramco and the Iraq consortium, IPC, were "puppets who danced at the will of their masters in New York and London," the Arabs said. "Hence, Tariki's demand for integrated operations subject to Saudi Arabian direction." Wanda "had done all she could to justify existing arrangements—mainly by pointing to the economic advantages which had accrued" to their countries through capital investment and considerable revenue. "But in almost every case, she was sooner or later faced with the poser that if the circumstances were reversed, Western countries would not tolerate the exploitation of their one and only asset by foreign companies controlled from foreign capitals."[26]

Wanda's conclusion hit hard: the existing joint ventures in the Middle East, exemplified by Aramco and IPC, would be "short-lived," she told Jersey's board. The only solution was free competition among the companies. And what then? Wanda "makes much of the point that the handling of oil company affairs in the Middle East contrasts strongly with practices in Venezuela," Williams wrote in a letter to a colleague.

"There, the major companies have traditionally pursued a policy of growing decentralization, with control from the center the exception rather than the rule." What's more, Wanda warned Jersey, "Knowledgeable people such as Tariki are fully aware of this difference, and make the most of it."

To her surprise, her main adversary on the board was not its senior Middle East expert Howard Page, as she had expected, but Jack Rathbone, Jersey's president. Though she admired Page's intelligence and negotiating skill, she had often found herself at odds with him over oil nationalism. But she had not dealt much with Rathbone, the nononsense engineer from Jersey's Baton Rouge, Louisiana, refinery who knew little about the international scene. He had just returned from a three-day tour of the Middle East with an optimistic picture. Rathbone "charged Wanda with excessive pessimism and she, in turn, told him that he had got his impressions from the wrong people!" Williams wrote. Daniel Yergin also recorded her recollection of the encounter: "'You never got beneath the surface,' Wanda told a stunned Rathbone. 'Jack, do yourself a favor. You got the red carpet treatment, you were there only a few days. You'd be wiser if you didn't make those statements.'"[27]

Although Rathbone did not show her the door, Williams said, she was "gloomy" about her mediation efforts. She then talked with presidents and board members from Aramco's other parent companies, who welcomed her "frankness" but expressed dismay at her message because the "Aramco and IPC empires can't be carved up at the stroke of a pen." Indeed, the fundamental policies of both Aramco and IPC remained unmoved: they would resist any efforts, whether from governments, other companies, or the press, to restrict their control.[28]

The oilmen, however, could not stop what many saw as Wanda's lobbying on Tariki's behalf. No matter how well she understood the industry's point of view, she continued to press her case. The companies should acknowledge the importance of oil nationalism and take steps to share control. In late August 1958, she sent several American and British chief executives a scathing, but unsigned, Saudi memo infused with oil nationalism. Jack Rathbone wrote to say that he needed to speak with her about it. Howard Page thanked her for his copy and said it sounded like Tariki, but dismissed it, saying, "It is difficult to separate the emotional from the intellectual and the facts from wishful thinking." John Loudon, Shell's managing director, was more attuned to the long-term implications: the memo made "a number of points the oil companies must take to heart if we wish to maintain our position." Mobil president Albert Nickerson made an "impressive" list of "all the aims and objectives of the Arab Nationalists" stated in the memo, and then concluded to Wanda: "They also fit rather closely with views that you have previously expressed." Despite these acknowledgments, the majors would not budge. Their chief executives "felt very threatened by Tariki," Marashian later said, but few would admit it.[29]

Petroleum Week's continued coverage from the Middle East made them wince, too. Both Wanda and Marashian regularly turned up scoops, from news about a rapprochement between Saudi Arabia and Iraq in 1956 to details about Saudi and Kuwaiti deals with Japanese companies and the largest U.S. non-major, Standard Oil of Indiana (Amoco). Particularly unnerving for Aramco's owners was the article that spelled out the terms of the Indiana concession. Marashian had sent the contract to Wanda and Macris, noting that this was the first copy of the agreement in the hands of anyone outside of the negotiators. He then added, with implicit reference to MacPherson, "A kindly person, u no who [sic], gave me a carbon copy." MacPherson was not the only friend of Wanda's who helped. She also got tips from Ed Brown, the Jeddah representative for Getty Oil, another independent company on the lookout for a Middle East deal. Wanda typically got more "inside dope" on Aramco from its competitors than from its parent companies.[30]

By 1958, Wanda's mere presence in the Middle East could set off alarm bells. Pan American, the international subsidiary of Standard Oil of Indiana, "was in a panic because one of their men walked into Tariki's office in Riyadh to find her sitting there," Macris wrote with delight to Marashian. Company executives contacted *Petroleum Week's* editor, LeRoy Menzing, to "beg him not to run any story" for a few weeks for fear of ruining their bargaining position. In fact, Wanda had not sent anything yet, but she soon did.[31]

Wanda continued to make her presence known outside the international oil world. In April 1958, she received another Jesse H. Neal award for her 1956–57 series of columns. This time the ceremony was at the National Press Club—where she would not have been relegated to the balcony—but she did not attend since she was on a Middle East trip. Despite her stage fright, she agreed to speak on radio shows and at conferences. She shared the podium, for instance, with CBS News' Walter Cronkite and Mobil president Albert Nickerson at a New York conference for Oil Progress Week at the Waldorf-Astoria Hotel, where she managed "in a very ladylike way," in the words of one oilman, to remind the audience "that there was something larger than Texas." To support her contention that oil reserves in the Middle East dwarfed those in the United States, she gave a colorful example that the *New York Times* repeated in its coverage. When visiting the sheikhdom of Kuwait, she was told by an oilman that "if a crew brings in a well that yields less than 5,000 barrels a day, they send them home to Texas." Of course, the *maximum* production allowed at most Texas wells, she added, was only five hundred barrels a day. The newspaper that would not even grant her a job interview in 1954 found her eminently quotable only a few years later.[32]

THE CARACAS COUP

In between her two big trips to the Middle East in 1957 and 1958, Wanda also fit in a seven-week expedition to Venezuela that led to another acclaimed series of articles. She arrived in early December 1957, just in time to watch the collapse of the military dictatorship that had dominated the country for a decade. Venezuela was still the world's largest exporter of oil, with daily production double that of Kuwait and triple that of Saudi Arabia. From 1948 to 1958, Venezuela's military junta pampered foreign oil companies with lower taxes, new concessions, and the suppression of unions. Its commitment to the American struggle against communism added to its stature in Washington, but the dictatorship's excesses were about to undo the regime.[33]

Her visit started out rather breezily. Before plunging into meetings, dinners, and field visits, Wanda agreed to be interviewed herself—in a Caracas amusement park. Ruth Pearson, the wife of McGraw-Hill's

Caracas correspondent, John Pearson, and herself a reporter, took Wanda for an afternoon of popcorn, Ferris wheel rides with the Pearsons' baby daughter, and lots of questions about the Middle East. "This is better than riding sand dunes in the Rub-al-Khali!" Wanda told her host.

Ruth Pearson was somewhat awed by Wanda. "A small, dark-haired, dark-eyed woman with a piquant face and an air of New York chic that belies her adventurous job," Ruth Pearson reported in the English-language Caracas *Daily Journal*, "she admits that being an oil editor surprises her as much as it does others."[34] She explained years later, "Wanda, in a way, was a role model. She was a pioneer in business journalism. None of the men covering oil ever came near to what she could do." Pearson heard criticism from men that Wanda was too assertive, but she disagreed: "Wanda was honestly outspoken—not aggressively so." Trading information was part of the reporter's game, but since Wanda personally knew all the chief executives and the oil ministers, she worked "in a different stratum—she had her own set of rules." She also knew how to get people to talk: "Wanda had this way of looking at a person as though he were the only person in the world worth talking to. She had that kind of intensity." Other Venezuelan papers also featured Wanda, but in a more formal way, citing "this notable authority" on world oil, whose "clear insight into this 'man's world'" allowed her to excel "in a most unusual field for a woman."[35]

John Pearson, later *Business Week*'s Latin America editor, also recalled that Wanda's connections immediately gave her access and insight that he and other reporters did not have. At lunch at the Hotel Tamanaco, overlooking Caracas, a few days after her arrival, she told Pearson and *New York Times* correspondent Tad Szulc that she did not think the regime could survive, that "something was brewing." Pearson, who had been covering Caracas politics since 1956, disagreed; Szulc was noncommittal. "As it turned out, I was wrong and she was right," Pearson said. "Her sources were better than mine. She saw everyone of importance—all the top oil people. They had their own intelligence networks, but she was important to them, too."[36]

On New Year's Day 1958, the Venezuelan air force rebelled. Troops loyal to the regime initially suppressed the revolt, but after three weeks of demonstrations, strikes, and armed conflict left hundreds dead, another military faction prevailed, one committed to restoring civilian rule with democratic institutions. Despite the rioting, Wanda kept up a hectic pace of interviews. In *Petroleum Week*'s lead story on January 17, she reported that oil operations were not expected to be disrupted since the conflict centered on civil-liberty issues and political freedom, not oil policy. The majors had been improving their public image by making the industry more Venezuelan, even requiring their foreign staff to learn Spanish, she reported, and many more Venezuelans were being hired in managerial positions.[37]

Her next article, however, drew considerable prepublication attention from U.S. officials. She had spent three hours interviewing the minister of mines and hydrocarbons, Edmundo Luongo Cabello, just after he survived sweeping cabinet changes. The American embassy's first secretary reported that the embassy "was able to come by a copy of a preliminary draft"—though heavily marked up—of a critical first look at the new junta's oil policy, and he made sure Washington heard about it right away. Venezuela was committed to "practical" nationalism, Luongo Cabello said. In a follow-up piece, Wanda described how well Venezuela's senior bureaucrats—all engineers and technicians with doctorates from universities in the United States—disproved the prevailing industry view that most government officials in oil-producing countries were incompetent and corrupt. "It doesn't do you any good," one U.S. executive told her, "to send a case of whiskey at Christmas."[38]

Wanda's several trips out of Caracas were a welcome relief from the turmoil in the city, although most of her destinations were rather remote. Getting out to Venezuela's first offshore drilling platform in the Gulf of Paria, near Trinidad, took more than six hours by plane and boat. She was intrigued that five American companies had gambled millions for the right to "go fishing" for undersea oil, where the fish were "definitely nibbling." The first test from their luxurious air-conditioned rig pumped out good-quality crude, but the operators were cagey. Details were closely guarded, she reported, because nearby exploration tracts had not yet been put up for bid. The Americans looked pleased with the results, even though they were cooped up on a rig "in a featureless expanse of blue Caribbean water." Growled one Oklahoman, "No rum, no beer, no senoritas."[39]

As much as she could, Wanda still tried to combine substantive coverage with local color. Along with the latest news from a huge refinery, she described its lonely location on a "flat, parched, and monotonous" peninsula where "a West Texan would feel at home amid the mesquite and cactus." The refinery's noteworthy features: thousands of goats that chewed up cable lines and an incessant wind that "makes one's hair stand on end and gives people a permanently surprised look."[40] Lake Maracaibo, a large shallow lake full of drilling rigs, proved more exciting. A spectacular discovery, she reported, had drawn a flock of "wildcatters"—small companies willing to drill in a generally unproven area. "There are more Texans (plus a sprinkling of Oklahomans and Californians) per square foot at Maracaibo's ritzy Hotel de Lago than any place this side of Houston," she wrote. But this was "a blue chip game. It's not for the timid or light of pocket." The first dry holes "cost a cool $1 million a piece."[41]

From Barinas, another hot exploration region, she wrote, "Getting boiled alive in oil hasn't been a problem since Biblical days—but it is here in Venezuela's newest oil producing province." With crude gushing out at extreme temperatures because of a low gas-to-oil ratio, "if a man should get accidentally sprayed," the field boss told her, "he'd be french-fried." After a rain shower, workers "watch the water puddles around the flow lines boiling merrily away." This Barinas oil story, she wrote, "is a classic example of the big, calculated risks involved in oil exploration—the high costs, and the long time often required before even a successful venture can start paying out." Since childhood, Wanda had had ground-level views of the risks and rewards of the oil industry. These trips only confirmed them.[42] And so did her next venture—the Venezuelan jungle.

DEADLY ARROWS
Wanda's stories from her trip to Venezuela's interior began on a chilling note. A few weeks before she visited the rain forest, a wiry young Span-

ish surveyor, whose last name was Vila Mas, had set out from his camp
near the Colombian border with several oilmen and armed guards to
check out Shell Oil's new CR-4 discovery well. In a Dodge Power
Wagon covered with a heavy wire mesh, they drove down a new mud
road that was "slashed, like wide curving pink ribbon, through the
steaming black jungle," Wanda reported based on Vila Mas's account.
Several miles in, they screeched to a halt. Three tree trunks "lay myste-
riously across their path"—an ambush, whispered Vila Mas. The men
stepped out, guns drawn. "In that instant, a shower of 4½-ft.-long barbed
arrows shot out of the black thickets on either side of the road, narrowly
missing the men. The guards fired back at their unseen assailants. Another shower of arrows. Another volley of gun shots. Then silence. 'No
runs, no hits, no arrows,' cracked tool-pusher Ed O. Seabourn," one of
the oilmen she interviewed. Just the start of "another routine day" for oil
workers in western Venezeula.[43]

These were "the traditional hunting grounds of the most primitive,
savage, and hostile Indian tribe: the dread Motilones," Wanda explained
in an extensive article about the challenges that geologists and drillers
confronted in exploring for oil in the Amazonian jungle. Shell Oil and
its competitors faced armed resistance, even if the indigenous people
had only arrows to fling against men with guns and Power Wagons. But
the lure of finding more oil was strong: Shell's production from this region, though modest, was some of Venezuela's best quality—an easily refined "sweet" crude. Shell took some risk in allowing a journalist to visit
one of their drilling camps deep in the jungle, but Wanda, once again,
was persuasive. It took an hour for a bush pilot to fly her from Lake
Maracaibo to Shell's temporary trailer camp for a crew of about a hundred men that was hacked out of the virgin forest near the Rio de Oro.

Very little was known about the Motilone native people, who also
harassed oil explorers in Colombia, she wrote. No Westerner had been
able to establish a dialogue with them, despite many attempts. The most
recent, she learned, "was by a missionary, who spent a long period 'soft-
ening them up,' by dropping parcels of food, cloth, machetes, and so
forth, all wrapped with a picture of his face. Then he went in. The missionary was not heard of again, until his head was found, cut off with
one of his own machetes."

Two days and a night at base camp gave Wanda a taste of how the crew worked, knowing their every move was "watched by unseen, silent, hostile jungle eyes." Nighttime was the most disconcerting: "Then the place takes on an eerie aspect." Despite floodlights and guards, the Motilones sneaked into camp, primarily to pilfer. The crew showed her the natives' "tell-tale marks in the mud: the distinctive footprint with no instep, and a big toe that sticks way out to the side (apparently deformed by pulling down on the bow with the foot when firing arrows)." Although the Motilones stole many things—hammers, cans of paint, clothing, even the airplane windsock, they did not take food. "Can't say I blame them," chipped in one worker, with a meaningful scowl in the direction of the camp's Chinese cook.

The surveyors and road builders faced the greatest danger. Who went out first in the jungle? "Three dogs," Vila Mas told Wanda with a grin. "The Motilones have a pungent smell, which the dogs don't like." The surveyors followed with "machete men," guarded on either side in box formation as they hacked away at the dense vegetation. "Quite safe on the inside—but I'd sure hate to be one of the dogs," said Vila Mas. Shell's hard-bitten superintendent of operations, Charlie Brown, who initially insisted he "gol darned wasn't going to let any old Indian scare a Texan," quickly gave in. He did not install wire mesh on his Dodge wagon until the day "a Motilone arrow plunged without warning into the side of his car, missing him by inches." During her visit, Wanda and her escort came upon an abandoned camp clearing where they found monkey bones from a recent Motilone dinner. There had been no recent deaths among the oilmen, she reported. The last serious casualty was two years earlier, when an arrow sliced right through one worker. The night she stayed at the jungle camp, however, "the Motilones attacked an isolated farm not far from here on the Rio de Oro, reportedly killing four persons."

She was so fascinated by these native people that during her return flight to Maracaibo, she asked the pilot of the twin-engine plane to circle down inside a clearing to take a closer look at one of their communal, peaked-roof grass huts. They got quite close. "The Motilones I saw were reddish-skinned, nude except for G-strings, rough-looking, and stocky, with virtually no neck, and heavy-lidded eyes." Some even "climbed on their high roof to gape at us—and thus we had an unusually close-up view." When they started running for their bows and arrows, "we got out of there fast," she wrote, but she took photos to prove she was there. She also asked her hosts to take snapshots of her checking out a barbed arrow and listening to the oilmen. Dressed in slacks, a striped chemise, and pearl earrings, she appeared remarkably elegant and at ease among these tough men. Back in Caracas a few days later, she was the envy of all the foreign reporters. No one else had been invited to the jungle, said John Pearson, although they would have "jumped at the chance." No one else had that kind of clout.[44]

Wanda's columns from Venezuela echoed themes from her Middle East coverage. The challenges for oil geologists and drillers intrigued her; she identified with these hardy men who, like her father, thrived on the physical hardships and mental challenges of the hunt for black gold. Because she appreciated their work and understood their business, she could coax tight-lipped oilmen, typically suspicious of reporters, into telling her far more than they had planned. She, too, had lived in oil camps, she would tell them. As a child, she had tagged along with surveying expeditions and even fallen into an oil pit. This ingrained knowledge of the physical and financial risks made a difference. Her reporting humanized the industry for readers unfamiliar with the harshness of field work and the complexity of major oil investments.

Another basic tenet for her was a more common postwar belief that American-style modernization and industrialization improved conditions in less developed countries. The "black jungle walls" of the rain forest were to be penetrated, not preserved. Her report on the Motilones, while almost anthropological in its detail, still labeled them "savage." Her descriptions of refinery expansions and oil-town booms reflected her belief in progress through Westernization and prosperity. Communities that encouraged home ownership, instead of the paternalism of company towns, were part of the oil industry's "big and imaginative experiment" to encourage a free-enterprise society. She concurred with the oil minister, Luongo Cabello, when he said the oil companies' prosperity "creates more prosperity for everyone else in Venezuela."[45]

She had also reported on the benefits of modernization to the Middle East thanks to the petroleum boom, but there she showed a different sensitivity. The Bedouin, though illiterate, were not savages or "ragheads," as some oilmen called them, but were people of "real character." To her, the Bedouin represented the native intelligence and nobility of an ancient people. For Wanda, despite her passing fascination with the Motilones, the South American rain forest held none of the romance of the Arabian desert.[46]

COLD WAR COMPLEXITIES

For most of the 1950s, the majors remained masters of the world oil scene. By presenting a united front and capitalizing on the support of Washington and London, they were able to humiliate Mossadegh and to demoralize Iranian and Arab nationalists alike. Despite genuine rivalries, the seven majors shared common goals through interlocking joint ventures in the four prime Middle East oil-producing states. These partnerships—Aramco in Saudi Arabia, IPC in Iraq, Kuwait Oil Co., and the Iran consortium—served to restrain internecine competition. They developed intricate rules for overall production rates and each company's proportionate share of costs, operating rights, and "offtake" —the number of barrels allowed per owner. In effect, the world oil industry became what economist Edith Penrose, a longtime friend of Wanda's and a professor at the London School of Economics, called a "partial oligopoly" to describe a situation in which a handful of sellers with common interests controlled almost the entire market. Despite new supplies from the Soviet Union, the majors produced and shipped most of the world's petroleum through their vast network of affiliates. It was still, as journalist Anthony Sampson noted, "a charmed circle."[47]

But by the last years of the decade, the majors' postwar petroleum order was beginning to fray. The spread of Nasser's nationalistic fervor, the Soviet Union's support for this struggle against the West, and the threat to the seemingly sacrosanct fifty-fifty oil deals from a wide range of competitors began to undermine, at least a little, this de facto oligopoly. So, too, did a perennial problem for the petroleum industry—supply and demand. The extraordinary surge in worldwide oil consumption during the 1950s was no match for the tidal wave of discoveries. This exponential growth seemed to dwarf previous boom-and-bust cycles. World demand was expanding quickly, but crude production nearly doubled over ten years, and reserves skyrocketed. What's more, stepped-up Soviet production led to higher exports at cut-rate prices. Although less than a fifth of supplies were sold on the market—the rest were sold under long-term contracts—the majors came under increasing pressure to discount their official posted prices. Yet they paid royalties and taxes to oil-producing countries based on these posted prices, which were linked to U.S. crude prices: these in turn were artificially inflated because of the partially regulated import ceilings set by state commissions (the partial regulation that got Wanda into trouble at the Senate hearings in 1953). Because of the glut in the late 1950s, every time the majors sold crude at a discount from the posted prices, they alone absorbed the loss.[48]

Iraq's bloody coup in July 1958 refocused Wanda's attention on Soviet designs on the Middle East and the need to keep the region within the West's sphere of influence. In an explosive uprising, stirred in part by Nasser's three-year war of words against the British-backed monarchy, Iraq's army revolted. Supported by crowds carrying photos of Nasser, troops killed the king, the crown prince, and the prime minister. "Western control of the Middle East's vast—and vital—oil reserves was visibly shaken," *Petroleum Week* reported that week after details of the grisly assassinations came out.[49]

Though Nasser's nationalism both worried and fascinated Wanda, her anticommunism was unwavering. Her conviction that the West had to defend its oil interests in the Middle East led to a number of columns correcting Western misconceptions about the region and advocating a more "enlightened" industry approach. A daily dose of Soviet propaganda "on Moscow's booming Arabic language broadcasts," she wrote, was conditioning Arabs to look at any American action as imperialistic, as an "oil grab." Arabs were being "brainwashed." The label "Middle East oil" tended to be misunderstood by many Westerners, as it suggested that the region's vast reserves were "bubbling in the same political pot." Iranians were not Arabs, and, given their history of struggling for independence against Russian and Mongol invaders, they were proud of their Aryan heritage and "tend to regard the Arabs with little concealed scorn." Many Iranians viewed "Nasserism" with "open alarm." What's more, during Iran's nationalistic bid under Mossadegh, its Arab neighbors did not protest the majors' boycott—instead, they benefited from Iran's shutdown. Western oil planners, Wanda advised in August 1958, should actually build up Iran's production capacity "as potential insurance against Arab oil crises." If Iranian oil was not the West's trump card, "it is at least an ace."[50]

The shah, she soon learned, was pleased. "His Majesty praised your breadth of vision and perspicacity of your comments," wrote Fuad Rouhani, vice chairman of Iran's national oil company, the NIOC. Her old friend Hussein Ala, the former prime minister and then minister of court (again), appreciated that she understood Iran's "special position and problems." This mistaken association of Iran with the Arab countries, he wrote, "causes many misunderstandings and misinterpretations" with Westerners. Farmanfarmaian, the suave NIOC official who found her so engaging during her 1953–54 trip, was also pleased. "Taking this pen I feel like writing to a sweetheart well known to me," he wrote in elegant script. He was more than a little frustrated by the West's lack of sensitivity to Iran, let alone the fact that the majors had boosted Iraqi output to near parity with Iran. "We have to rely," he added, "on the kindness of a charming woman and your influence to tell these people that we are still badly treated."[51]

The role of the oil majors, Wanda wrote in *Petroleum Week* a week later, was also misunderstood. The flexibility that these companies gave to consuming countries, through their investments in a number of oil regions, was "an indispensable form of supply insurance." Direct, country-to-country oil deals, she argued, were fraught with risk; oil companies provided a buffer. Price was important, but continuity of supply was even more so. Despite her growing sympathy for the oil countries, she believed the majors served a vital function.[52]

Wanda identified destabilizing factors in addition to those posed by the Russian threat to the status quo. "Watch for a brand new trend" that could revolutionize the oil industry worldwide, she warned in a 1958 column. The majors, though still dominant, were about to be attacked from their own rear flank. Frustrated by limited growth potential at

home, several more U.S. domestic companies—"the independents"—
were about to plunge into the more lucrative international market, from
wellhead to gas pump. As one company president told her, "We'd rather
share the headaches of big growth than sit back home comfortably
while others pass us by."[53]

For Wanda, those headaches—those threats to the status quo—
meant more news. On Iraq, she had already irritated Geoffrey Herridge,
IPC's managing director, by publishing many details of IPC's operations and giving extensive coverage to Iraqi oil officials. And when, in
1958, a source slipped her a copy of an embarrassing IPC internal handbook that cast doubt on Oman's sovereignty and the company tried to
recall every copy, Wanda did not comply.[54] When Iraq's new pro-Soviet
regime demanded a major overhaul of IPC's concession soon after its
1958 coup, Wanda again foiled Herridge's efforts to keep things quiet.
On the surface, her December 1958 column on the Iraq negotiations
appeared rather favorable toward IPC. She praised the company for relinquishing a swath of choice acreage in central Iraq, a move "unique in
the annals of concession operations" because "nobody—but nobody—
in his right mind in the oil business ever voluntarily relinquishes any
but the least promising areas of his concessions." The IPC decision
showed "an enlightened degree" of flexibility and openness "for changing with the times."

But to anyone familiar with the scene, her irony was obvious. Packed
between seemingly favorable paragraphs were details of IPC's several
offers and counteroffers—details that Herridge had sought hard to keep
secret. The numbers told the story: Iraq was successfully browbeating
IPC into giving up at least 50 percent of its vast concession territory.
IPC's London-based managers, known as the most intractable among
the majors, were being forced into "flexibility." Wanda's damning with
faint praise did not go over well with IPC. Two months later, in a letter
to Onnic Marashian, who was about to relocate to Britain, Wanda gave
him some advice with a tongue-in-cheek reference to the quality of
crude oil, known as "gravity": "Have fun in London—and when you go
near IPC, pretend you never heard of me—Herridge wants my head
boiled alive in oil, and he is not particular about the gravity."[55]

A PRICE CUT

In February 1959, under mounting pressure from the oil glut's financial drain, the majors took what seemed to be a reasonable step: they cut the price of crude. British Petroleum was the first to move, slashing posted rates by 10 percent—to about $1.90 a barrel. Eager to maintain market share, the other majors fell into line.[56]

The oil-producing governments were stunned. Accustomed to steadily increasing revenues from years of expanding exports and stable prices, they had to face an immediate slash in income. What's more, they had no advance notice. Abdullah Tariki was outraged, as was Venezuela's newly appointed oil minister, Juan Pablo Perez Alfonso.[57]

For a decade, Perez Alfonso had lived in exile. He had taken refuge in Washington, D.C., after the 1948 coup that ousted Venezuela's first democratically elected government. Eking out a living by writing articles, the scholarly refugee spent hours at the Library of Congress reading oil journals and learning how the Texas Railroad Commission in the 1930s reined in the glut through production controls. In 1954, he moved to Mexico City, where the rent was cheaper, but he sorely missed the library. When Wanda learned about his difficulties in 1956, she arranged for him to get free copies of *Petroleum Week*. "I thank you a million times for your kindness," he wrote her. The journal's information was "very complete"—a lifeline keeping him up to date. When Venezuela's dictatorship collapsed in 1958 and Betancourt returned to power through national elections, the new president asked Perez Alfonso to take charge again of the Ministry of Mines and Hydrocarbons.[58]

But the oil world had changed in ten years. Although Venezuelan production had doubled, the exponential growth in Middle East exports posed a huge threat. Venezuela's production costs were much higher, and lower prices not only meant lower income but also lower investment in future development. Concerned about the decline of "the national heritage," Perez Alfonso also wanted to find ways to conserve oil for later generations.[59]

The following month brought another shock. The February price cut drew a deluge of complaints from the politically powerful U.S. independent producers, now suffering from a flood of imports that un-

dermined domestic prices. So in March 1959, the Eisenhower administration replaced voluntary import restrictions with mandatory quotas.
Mexico and Canada were exempt for security reasons, but Venezuela,
the United States' largest foreign supplier, was not, underscoring Venezuela's vulnerability to the surge in cheap Middle East exports. Its
boom could easily go bust. Perez Alfonso decided to move quickly—to
Cairo in April and Washington in May. He wanted to find a way to limit
production enough to keep supply in balance with demand and prevent
further pressure on prices. Cairo was the venue for the first Arab Petroleum Congress, a long-awaited gathering of representatives from a
number of Arab states, sponsored by the Arab League. Although Perez
Alfonso came as an observer, he had plans for action. Venezuela, he announced before leaving for Cairo, should form a "common front" with
oil-producing countries in the Middle East "to find means of collaborating to avoid arbitrary fixing of prices."[60]

MATCHMAKER
Wanda had told Tariki about Perez Alfonso. During their long sessions
in April 1958, Wanda had listened to Tariki's passionate plans for Saudi
Arabia to gain control of its oil industry bit by bit. He wanted not only
to press Aramco for greater participation, but also to create a national
oil company, seek foreign marketing experience, and—off the record—
coordinate efforts with other oil-producing states. To Wanda, this ambition sounded familiar. "There's another guy who's just as nuts as you
are," she told him. Who? Juan Pablo Perez Alfonso, the architect of
Venezuela's fifty-fifty deal and the most substantive thinker she knew
from any oil-producing country. Tariki had attended a conference in
Venezuela in 1951 but had not met him. "Where is he?" Tariki asked.
In exile, she replied.[61]

    She knew, from his letters to her, that the Venezuelan was still pas-
sionate about oil. In one letter, he had railed against the military dictatorship that was "throwing away . . . our best oil reserves." In another, he
had criticized the foreign oil companies: "Have you noticed how right
we were when we established the policy of NO MORE CONCES-

SIONS? Without an extra acre, they are suctioning twice as much oil than they were in 1948." *Petroleum Week*, he told her, was accurate; the higher output masked the fact that Venezuela was earning less per barrel than before. He was but biding his time in Mexico City, "preparing to be more useful to our country when the time arrives," he vowed. "It shall come, sooner or later." And by 1959 it had.[62]

Tariki cabled her, Wanda later recalled, after he had read a report on the newly appointed Venezuelan oil minister: "Is that the man you told me about? The nut?" Tariki "thought his name was Mr. Pablo," she later said, because she had referred to him as Juan Pablo. "Damn it, yes," came her reply. "I'll introduce you at the Arab Oil Congress."[63]

Wanda made her way to Cairo via Beirut, arriving early on April 13. At the last minute, she accepted a ride on one of Aramco's "milk run" flights from New York. Aramco's Jack Hayes, on his way to Dhahran, remembered distinctly that "Wanda and a vice-president were the ones who got the two sleeping berths" while the other passengers suffered in stiff seats. When she arrived, she woke up her father's old friend, Brandon Grove. She could not get a hotel room: Could he help? Grove had breakfast with her and let her use his room while he worked at his office, he noted in his diary. By afternoon, he had found her lodging for the night.[64]

In Cairo the next day, Wanda met up with Perez Alfonso at the new Hilton, where many conference participants were staying. Spotting Abdullah Tariki sitting in the lobby, she brought Perez Alfonso over to meet him. "So you're the one I've been hearing so many things about," Perez Alfonso told Tariki. She invited both men up to her suite for drinks—Coke or bourbon, depending on her audience for the story—so they could talk privately. She had just become, according to Daniel Yergin, "the matchmaker of an alliance that would develop into the Organization of Petroleum Exporting Countries—OPEC."[65]

"Did you have to?" financial writer Adam Smith asked her in a 1979 interview when OPEC seemed to have the entire world economy in its grip. By then, wrote Smith, OPEC had engineered the greatest transfer of wealth in the history of the world. "It seems to me that this is a bit like saying that you were the one who gave Admiral Yamamoto the guided tour of Pearl Harbor back before he had that idea," Smith suggested.

"Oh," Wanda blithely replied, "it would have happened anyway."[66] Although OPEC would have certainly challenged the Seven Sisters without her, Wanda's offhand response belied the role she had played in bringing its two principal founders to the fore and the significance of her influence over the global oil business in the coming decades.

6.

OPEC's
Midwife

Ultimately, some Western observers dismissed the first Arab Petroleum Congress—the oft-postponed gathering of Arab oil delegates that was finally held in Cairo in April 1959—as inconsequential. The expected firestorm of protest against the majors, particularly from Abdullah Tariki, did not materialize. No one issued a call for Arab states to nationalize their oil industries. Instead, "ignorance of what to Western minds are the elementary facts about the oil industry was the main feature of the Congress," British Petroleum's Michael Hubbard, one of many oil executive observers at the conference, noted smugly in a confidential report to his chairman.

Wanda, whom Hubbard called one of the "outstanding personalities" at the congress, disagreed with his assessment. Inexperienced and naïve many delegates at the congress might be, but some were quite savvy. She had seen British-trained Iranian engineers skillfully manage the Abadan refinery without BP in 1954, and although Tariki was hot-headed and idealistic, he had a quick mind. Hubbard's views exasperated her because his arrogance was so typical of many Western oil executives.[1]

Nonetheless, when Hubbard asked her for an introduction to the Saudi oil director at the conference, she obliged. She wanted both sides to better understand each other, and besides, it might lead to a story, or at least some inside information. "Wishing to find out privately the attitude of Tariki," Hubbard noted, "a meeting was arranged through the auspices of Wanda Jablonski, who claimed that from personal knowledge it was possible to discuss economic facts." Unfortunately, "this proved not to be the case and we were subject to a diatribe" on the inequities of Middle East production policies. In the end, "it proved quite impossible to establish any point of contact."[2]

The degree of ignorance on display in Cairo depended on one's perspective. Hubbard and his BP colleague A. T. Chisholm proved oblivious to both the public and the private significance of the Cairo gathering. To them, the papers presented were "worthless," "ill-informed," or "based on emotional concepts." Chisholm's meeting with one of Nasser's key advisors, Mourad Ghaleb, "led, as I expected, to a long, polite and vehement" denunciation of British policy, he said. Well briefed by the British Foreign Office before flying to Cairo, Chisholm made "successful" counterarguments. He also credited Hubbard's "adroit handling" of press questions with defusing another delicate public relations issue: Perez Alfonso's frequent press conferences criticizing BP for starting the industry-wide price cutback in February without advance notice, consultation, or even a courtesy warning to the oil-producing states. In the end, Chisholm was pleased with the "uncontroversial nature of the Congress, which in my view can be marked off as a 'plus' for the oil industry's future relations with the Arab host countries."[3]

Needless to say, Wanda's assessment was completely different. To her, the weeklong congress was significant as the first substantive effort by Arabs "to think collectively about oil." Although anger at the unilateral price cut ran deep, most delegates recognized their dependence on Western oil outlets. Wanda was convinced that their nationalism would bring far-reaching changes to the world of oil, and she saw Tariki and Perez Alfonso as persuasive leaders. However strongly she identified with the major oil companies, particularly the bravado and genius of their geologists and engineers, she now believed equally strongly that they were wrong to resist the nationalists' demands for greater participation in the industry.

Though she slowed down enough to attend some speeches, she was primarily, as Hubbard noted, "active behind the scenes," listening and probing. Indeed, she was more active than he could have imagined: she had picked up a whiff of something more, something secret between Perez Alfonso and his newfound friend Tariki, something about which BP knew nothing.[4]

POWER POLITICS

Gamal Abdel Nasser was still in his prime in 1959. Despite great tension, if not downright hostility, between certain Arab states, Nasser was the undisputed champion of the Arab world, the hero of Suez. Egypt's battlefield losses notwithstanding, he had emerged the symbolic victor, the first modern Arab leader to strike a blow at the European powers that for so long had dominated the region. He not only had nationalized the Suez Canal, but had managed to draw substantial foreign aid from both West and East to help industrialize his agrarian nation: positive neutralism, he called it. For a time, he skillfully played power politics.

Nasser also had broader ambitions. Pan-Arabism, he promised, would overthrow both political and economic oppression. His impassioned speeches, heard all over the Arab world on the new, high-powered Radio Cairo, regularly denounced Western imperialism and the creation of Israel. Nasser wanted Arab states to join together to use oil against the financial clout of the West. Undeterred by Egypt's own lack of petroleum, he was expected, at the Cairo congress, to try to persuade his oil-rich neighbors to use petroleum as a weapon to help all Arab states, the oil haves and have-nots.[5]

The majors, watching Nasser with growing unease, raised the alarm in Washington. In the early years of the cold war, the policy of containment had led Washington to support any kind of nationalism that could constrain Soviet expansionist designs. During the Eisenhower years, however, this strategy evolved, gradually becoming a policy of mistrust — if not outright opposition — toward nationalist movements suspected of communist or even socialist leanings. Secretary of State Dulles was particularly outspoken: "International Communism is on the prowl" to entice nationalist leaders, he declared in the wake of the 1956 Suez crisis. The launch of the Soviet Union's first space satellite, Sputnik, in October 1957 shocked many Americans, making them feel suddenly vulnerable to Russian missiles, and reports of a growing Soviet presence in the Middle East took on even more ominous tones. As a result, initial support for Nasser as a modern nationalist turned into suspicion for some U.S. policymakers, deep distrust for others.[6]

Increasingly attuned to geopolitical concerns, the Eisenhower administration showed itself willing to intervene militarily in the Middle East. Armed conflict between pro-Western and pro-Nasser groups in Beirut in 1958 led to a U.S. Marines expedition to restore order, but also to Arab denunciations of Western imperialism. By then, an anti-Nasser consensus had coalesced in Washington. Dulles saw the Egyptian president as a threat to America's principal interests in the Middle East— Israel, oil, and containing Soviet expansion. Although Eisenhower worried about how to "get ourselves to the point where the Arabs will not be hostile to us," Dulles, who had once called Nasser a bulwark against communism, demurred: "Nasser, like Hitler before him, has the power to excite emotions and enthusiasm," he told Eisenhower. For Dulles, Soviet domination of the entire region was already under way with the violent coup in Iraq: "The real authority behind the Government of Iraq was being exercised by Nasser, and behind Nasser by the USSR." Nationalism could be the first step to communist control of the biggest oil pools in the world.[7]

This vilification of Nasser resonated with many Americans doing business in the Middle East. In an influential 1958 article, *Fortune* warned about Nasser's "divine frenzy" to make Egypt powerful for the first time since the pharaohs. Like Dulles, *Fortune* found Nazi parallels: Hitler had conquered the emotions of his own countrymen, but Nasser was conquering the emotions of an entire region. Even as Göring "used to cow Germany's neighbors with displays of the Luftwaffe," Nasser bragged about Egypt's industrial progress since it had shaken off Britain's imperial yoke. Yet Nasser was proving to be a dangerous demagogue "who understands no more about economics than most political prima donnas," but had "a messianic determination to raise Egyptian living standards." Most of his "placid" countrymen were uninterested in his lofty goals, but Egypt's intellectuals were enthused. At night "in his chamber," *Fortune* concluded, "Nasser no doubt exults in his progress."[8]

Some CIA officers disagreed. Kermit ("Kim") Roosevelt, who had organized the 1953 coup against Mossadegh, and Miles Copeland and James Eichelberger—all contacts of Wanda's—argued that Nasser's real goal, aside from power, was independence from foreign domination. To them, Nasser was no Soviet puppet, yet he could damage U.S. interests in other ways. An ardent nationalist and populist who had suffered under the British, he wanted to improve the lives of Egypt's poorest, but to stay in power, the CIA officers believed, he advocated a strong government with police-state tactics and might encourage subversive activities or violence to bring reluctant neighbors into the Pan-Arab movement and threaten Israel. Unaware of this more nuanced view, the American press generally disparaged Nasser in 1959, making Wanda's coverage from Cairo all the more surprising.[9]

NO FIREWORKS

The Egyptian president startled many delegates on the opening day of the Cairo conference by driving up to the exhibition hall in a well-worn Cadillac, stepping out without assistance, and stopping to light a cigarette himself. "This carefully choreographed performance was designed to show us that he was not a dictator but a man of the people," recalled Farmanfarmaian. The Iranian aristocrat, attending the congress as an observer, was surprised when Nasser greeted him with a friendly handshake and toothy smile, since Iranians and Arabs had a long history of rivalry. Indeed, Nasser chatted with oil-company officials as well.[10] This low-key opening set the tone for the congress. "Absent were the expected fireworks, soapbox oratory, and demands and recriminations against Western oil companies that marked so many of the advance Arab statements on the subject," Wanda reported in her first column from Cairo.

Absent, too, from the formal program was Perez Alfonso's much-publicized proposal that Middle East producers limit production in order to stabilize prices. The only serious controversy came from Saudi Arabia's American lawyer, Frank Hendryx, who spelled out a legal argument for the sovereign right of countries to revise their contracts. The Arab reaction was split; some delegates supported it, whereas others preferred redress through arbitration. Most of the papers presented, in fact, were technical or economic. "If it wasn't for the romantic Nile setting and the Arabic language," quipped Wanda, "this could have been just another routine session" of an American oil convention. It was "a far cry" from the Arab League's original goal of developing a unified policy to increase Arab control of the oil industry, but, as she explained, heightened tensions between Egypt and Iraq had intervened. Although Iraq's 1958 left-wing coup did not immediately change the country's oil policies, the uncertainty that it created undermined Nasser's hopes for a unified Arab stance. By refusing to send an official delegation to the 1959 Cairo congress, Wanda wrote, Iraq "killed" Nasser's "punch line."[11]

American press coverage of the congress was, for the most part, superficial or sensational: Nasser was making a "red grab for riches"; "defiant" Arab delegates, trying to "lay down the law to the big oil companies," failed to do anything but talk; and Arabs were simply "out for more money and more control over their oil." Tariki did get some attention, two years after Wanda had called him "the No. 1 man to watch." According to *Time*, everyone at the conference wanted to see this "incorruptible" Saudi "with a bright, quick smile, and a profile as sharp as a scimitar," the unquestioned spokesman of a new generation of ambitious Arabs. The *New York Times*, without an on-the-spot reporter, presented only the majors' point of view, but the *Wall Street Journal*'s correspondent noted the possibility of closer cooperation between the Arabs and Venezuelans: "Nothing along this line was agreed to," but the idea was bound to return. One of the *Journal*'s editorials, however, was scathing: if Arabs tried to do what Mossadegh did in Iran, they, too, would learn, "to their regret," that "sensible men sign contracts only with people they can trust."[12]

Wanda, as usual, not only got exclusive interviews, but also provided greater insight. She explained that the Arab League's Mohammed Salman, a future Iraqi oil minister, urged the majors to "restudy" their agreements to rid them of unfair provisions—a more moderate request than a "demand" for contract revisions, which some journalists incorrectly reported. She also detected a shift in Tariki's position that others missed. His "favorite demand for sharing in overseas refining and marketing profits," she noted, "was touched on only in an academic way." The price cut, the oil glut, and the need to find common ground was leading the Arab nationalists to reexamine their priorities.[13] As it turned out, the Venezuelans picked up on this shift, too, and coaxed the Arab delegates into considering alternatives. They had, in fact, a radical idea of their own.

Exactly what this was she had yet to find out. Perez Alfonso and his colleagues were lobbying hard for some kind of production and price-stabilization arrangement among the oil-producing countries. Tariki found the idea appealing, but, at least publicly, he "admitted its practical realization was a long way off." The first step, he told Wanda in private, was to form a committee of oil-producing countries "to meet periodically to discuss mutual problems and to evolve a unified policy."[14]

With that, he had just given her a good tip, so she kept asking questions. Although Arab delegates were upset with the arbitrary 10 percent price cut, they widely recognized some critical facts: Arab oil had little value without Western markets; these markets were mostly controlled by the majors; Russia was more of a threat than a support; the glut was so great that the world could easily withstand export interruptions from one or two countries; and it would be "folly" to nationalize or try "too abrupt a use of pressure." Even the new Iraqi government, contrary to Western reports of imminent nationalization, was "behaving in a conspicuously correct manner" toward the oil companies.[15] To Wanda, the overarching theme of the congress was a demand for increasing Arab participation in the industry. "Naturally," she wrote, "they want more income from their oil, and equally naturally, they want a greater say." By speeding up the training of Arabs into positions of responsibility, the companies could increase their own stability.[16]

A pragmatist, not an idealist, Wanda had been making this point since her 1954 visit to the Abadan refinery. Certain oilmen in Cairo might agree, but she was writing for the decision-makers in New York, London, and San Francisco headquarters. Implicit in her column was a warning: the tone in Cairo was moderate, but it might not last unless the majors paid heed to these moderate men calling for a moderate level of participation in decisions that profoundly affected their countries. It was, as she said, only natural, but that was not what the captains of the oil industry wanted to read.

JEWELS IN CAIRO
Though most of the conference sessions were held at the Engineers Society building, the favorite rendezvous was the nearby Hilton, the ultra-modern hotel newly built on the banks of the Nile, where Wanda introduced Tariki to Perez Alfonso. This "turquoise jewel of Cairo," its brilliant, Mediterranean-blue tile exterior adorned with hieroglyphs, stood in sharp contrast to its surroundings—mostly nineteenth-century Italianate buildings in monochromatic shades of gray, laden with sand blown in from the desert. For Egyptians, the hotel's prized location also held special significance, recalled *Petroleum Week*'s Onnic Marashian. It was built on the site of "the ugly Kasr el-Nil barracks—the symbol of British colonialism in the heart of Cairo." It was in the Hilton, not at the conference, that Wanda spent most of her days—and nights. The most interesting talk, Wanda told her readers, came in the hotel hallways and lobbies, the private meals and cocktail receptions. The Venezuelans, particularly, were active, and she carefully watched their eager comings and goings.[17]

The official participants, plus a number of Western oil executives who had come as observers, hosted receptions on a seemingly endless cocktail circuit. Wanda's family friend Brandon Grove, Mobil's London chief, accompanied her to numerous parties, including a late-night dinner dance on a Nile cruise ship, the *Omar Khayyam*. They were joined by mutual friends Joe Ellender, a small, intense New Yorker who was a senior oil economist with Jersey, and George Ballou, the gregarious, bow-tied Middle East chief for Socal, and their wives. They "dined on *kabab*," Grove noted in his diary, and danced until midnight. Wanda and Grove, however, had to rise early because she had arranged for him to join her for breakfast with Tariki. Grove found this well worth the effort. The three of them spent more than two hours discussing oil issues, which "put me on a basis of solid acquaintanceship" with Tariki.[18]

Evening interludes that included a few women must have been a welcome change for Wanda after all the intense intelligence gathering she did with men. She got along well with the wives of many of her contacts, and several were friends she could shop with, something she enjoyed but did not like to do on her own. Though she genuinely welcomed their company, Georgia Macris said, Wanda's friendliness was also a bit calculated: it mitigated potential jealousy. "She could look at my wife and say with her eyes, 'I'm honest,'" explained the Kuwaiti official Feisal Mazidi, whose wife enjoyed entertaining her. "That was one of Wanda's secrets—she could be friends with men without offending their wives." Yamani's wife, Tammam, agreed, adding, "Wanda might have benefited from the fact that she was a small woman to stand out in a room full of men, but she did not flirt."[19]

Although she eagerly sought companionship, she was fundamentally shy. She did not like to be alone, Macris said. Wanda was afraid of empty time. Always visible, immediately recognizable in a crowd, she found it both an asset and a strain to be the lone professional woman in a sea of men—like the Hilton, a bright jewel in a field of monochrome. Only a few of her friends, such as Eleanor Schwartz and Georgia Macris, knew what an effort it was for her to attend the countless rounds of receptions. However serene she appeared to be, she had to take deep breaths to calm herself before walking into a party. To her shame, she also bit her nails to the quick.[20]

It helped to have an escort. Grove, whose ailing wife did not travel, often accompanied her at international gatherings. Another friend, New York oil analyst James Hunt, pretended on occasion to be Wanda's date for U.S. receptions. "It was tough to be a single woman in a room full of men," he recalled. "Wanda didn't want people to step over the line in a way she couldn't control." When she felt comfortable enough, she would give him a hand signal. *Petroleum Week*'s Paris correspondent, Helen Avati, had similar problems in France. At her first cocktail party as a reporter in the 1950s, all the other guests were men. "I walked in, and I walked right back out. It was terrifying," she recalled. "That's one of the biggest things that has changed since my early days."[21]

But the Cairo congress spoiled one of Wanda's friendships. Farmanfarmaian watched in growing dismay at the close attention she paid to Tariki: "Wanda stuck to him like a burr throughout the congress, making me feel neglected as one of her oldest friends," he wrote in his memoirs. "She was flirting only with Tariki," he later said, "even though she had known me so much longer." The Iranian admitted he was jealous: "I was hurt that she had given us up for the Arabs. She had taken their part." Tariki, with his dark chiseled features and broad smile, looked rather suave in a Western suit, but Farmanfarmaian called him ugly, particularly because he had one eye that tended to wander. Tariki also irritated the Iranian by "bragging endlessly about his swank new apartment" in Cairo's only skyscraper. But Farmanfarmaian would soon overcome his disdain.[22]

GENTLEMEN'S AGREEMENT

At a private dinner with Perez Alfonso during the congress, Farmanfarmaian learned about the Venezuelan's hopes for a confidential agreement to set up a regular dialogue among the leading oil-producing countries. Would the Iranian be willing to meet with a select group of Arabs and Venezuelans for further discussions? Farmanfarmaian agreed. It had to be "done in complete secrecy," Perez Alfonso warned. "No journalists. Not a word outside our little group." The following day, arriving in separate cars, the group met at a yacht club along the Nile in Maadi, a Cairo suburb, in what Farmanfarmaian called a "James Bond atmosphere." They talked outside under a solitary tree at a rickety metal table—Tariki; Perez Alfonso; Farmanfarmaian; Kuwait's top delegate; the Egyptian oil official who chaired the congress; and Salman, the Iraqi Arab League official who, like the Iranian, could not officially represent his country.

The "Gentlemen's Agreement" of Maadi, as it came to be known, urged the signatories' governments to establish a formal consultation commission as a way to defend against arbitrary decisions by the oil companies and improve concession terms for the oil-producing countries. Although vague and informal, the agreement was later seen as a milestone in the changing dynamics of the petroleum industry. It was the first significant move toward creating a common front against the oil companies.[23]

When he ran into Wanda later that evening, recalled Farmanfarmaian, she said pointedly, "You all came back from Maadi looking very secretive. Dear Manucher, do tell me what's going on." He remained mum; he was not going to let her coax anything out of him this time. Neither could she get Tariki to say anything—for the record.[24]

Perez Alfonso gave her a lengthy interview. He, too, was evasive, but Wanda made the most of what she got. He told her in greater detail than he did others about the special exemption he would propose in Washington after leaving the congress. He knew that he could alert U.S. officials to his request through *Petroleum Week*. When Wanda pressed him about a possible secret agreement, he alluded only to a "meeting of minds." Choosing his words carefully, Perez Alfonso said that he had reached no "concrete arrangement" but that "feelings" were "running in the same direction," particularly on prices. "Perhaps no formal agreement is necessary," he added, "if everyone works in the same direction."

Expert readers picked up the hint. Reporter Pierre Terzian later cited this quote as evidence that, despite all precautions, the American oil companies "had got wind of the existence of an agreement."[25] In London, BP's acerbic chairman, Maurice Bridgeman, must have been disturbed to read Wanda's column after digesting his own delegates' smug assessment of their public relations success. Readers attuned to the nuances knew Wanda was on to something.

Whatever she actually knew about the secret accord, she could see, in their demeanor and their words, that the Venezuelans were more than a little pleased by the results of the congress. When they first arrived in Cairo, they were anxious. Perez Alfonso had told her it would be tough to persuade these countries with deep-seated rivalries to consider a common position, to understand why they should get the companies to restrict production rather than expand it. The timing of the much-delayed congress, however, actually helped. As one Venezuelan said to her, if BP had cut the price after the congress, "the Arabs probably would have had no sympathy for our viewpoint." Although Perez Alfonso did not get the formal accord he had hoped for, the Gentlemen's Agreement was the next best thing. It was obvious to Wanda that Perez Alfonso was leaving Cairo a happy man.[26]

Her reporting diverged sharply from *Fortune*'s first in-depth assessment of Perez Alfonso and Tariki in its August 1959 issue, which explained their goals as an "ambitious scheme" to create an "international cartel." The plan was "not only impractical but bad economics." These countries could not come to an agreement and stick to it, *Fortune* postulated; besides, the majors still controlled the refining and distribution outlets. Ironically, these "self-styled oil technocrats" were modeling their plan on the Texas rationing system. Seemingly abandoned "as orphans who have been lavishly financed by rich foster parents" once the majors cut oil prices, they were now trying to fend for themselves. This assessment was soon obsolete, financial commentator Adam Smith wrote two decades later. The tone of *Fortune*'s article "was one of amusement and skepticism; its point of view, as usual, that of big business," Smith noted. "To read it is to enter an astonishing time warp." Wanda Jablonski, on the other hand, had correctly identified the nascent power of these nationalists' ideas.[27]

NASSER ON OIL

Wanda had other reasons to be pleased. Nasser finally agreed to give her an exclusive, hourlong interview—"the first public expression of his views on oil," she noted in the resulting article. Because he rarely spoke with Western journalists and had already put her off twice, this interview, Marashian said, was a "real coup." (Her contact with CIA operative Kim Roosevelt, then on good terms with Nasser, may also have helped.) Wanda found Nasser pragmatic and informed, with a "sound understanding" of oil fundamentals, not the unsophisticated demagogue he was often made out to be. He avoided interviews with Westerners, he explained, because whatever he said "would be immediately misinterpreted by the world press as attempts to control or grab the oil of the big Arab producing states." The president smiled, however, when she told him that U.S. correspondents were disappointed with the lack of political fireworks at the congress. "Westerners seem to expect Arabs to be entirely political in their approach to oil and other matters," he said. Under foreign domination, "our approach necessarily had to be political," but a postcolonial Egypt was free to focus on economic development. The oil congress provided a "constructive and needed interchange of ideas."[28]

Egypt's nationalization of the Suez Canal, Nasser said, should not be seen as a precedent for Middle East oil concessions. Arab countries understood that oil had to get to the customer before it had any value; they needed and wanted to cooperate with the West "as long as the terms are fair." Then Wanda interjected, as she was known to do when her hosts became long-winded, rephrasing his indirect answers with a blunt question. Was he saying "it is one thing for a state to nationalize a domestic facility, but an entirely different thing to try to nationalize an export industry, where the outlets are beyond the state's control"? Nasser said yes. Her signature move had drawn a significant statement.[29]

But she had more to report. What Nasser said about communism genuinely surprised her. He was worried that communists would gain control in Iraq and infiltrate Syria, not only endangering the entire Middle East but also isolating Iran, Pakistan, and India. The possibility of a communist takeover was "very serious," he warned. Given the prevailing American view of Nasser as a political prima donna under the influence, if not direct control, of Moscow, these warnings in May 1959 were ironic indeed. Wanda's interview with Nasser, printed alongside her column on Perez Alfonso, defied all expectations. She refrained from direct commentary, but the message was clear: Nasser opposed foreign domination of any kind, including communism.[30]

THE OIL GAME

Despite his success at the Cairo congress, Perez Alfonso, along with his new partner, Tariki, faced formidable foes. Their slingshot message, lobbed at the Goliath oil companies through the pages of *Petroleum Week*, barely made a dent. They were up against not only the combined clout of some of the world's wealthiest corporations but also the disparagement of the British and American governments and much of the Western press. In May 1959 in Washington, Perez Alfonso found he had no leverage; at a time of surplus, Venezuela's views held no weight. State Department officials listened politely and turned him away, and behind his back, oilmen derided his production-control plan. Even as savvy an oilman as John Loudon, the erudite Dutch-born chairman of Shell's board of directors, told British diplomats that Perez Alfonso was a "fourth-rate economist with ill-digested ideas."[31]

In 1959, the majors could still afford to be patronizing. Their preeminence, though challenged, remained strong, with minimal monitoring from Washington and London. Since ample supplies at steady prices were keeping consumers happy, U.S. and British officials felt vindicated in relying on the majors to maintain the status quo. Even the usual complaints from American domestic producers died down after they got mandatory import quotas to protect them from cheap foreign crude. Despite the glut and the ever-increasing number of independents vying for overseas deals, the majors retained effective control of international prices.[32]

The leaders of this club, a small circle of executives, continued to make the key decisions. In the rarified atmosphere of Jersey's Rockefeller Center or BP's Britannic House, said Jan Nasmyth, publisher of London-based *Petroleum Argus*, these men worked in "ivory towers," with a few key personalities dominating the boards. Secrecy remained the prevailing corporate culture. "There was a wall around you," explained Silvan Robinson, a Shell International Trading president. "There was just a small cabal of people who made decisions, either collectively or individually. But because of the company culture, they were actually very limited in their ability to get outside information." The U.S. majors, particularly, "were arrogant and naïve," according to Jack Sunderland, president of the independent company, Aminoil. Despite Wanda's reporting, Sunderland said, "they were fools about the massive changes under way." Although linked through joint ventures, the majors continued to jockey with each other for new production sources and greater market share. Yet when common threats arose, they drew together in common purpose.[33]

For legal and historical reasons, the European companies worked together differently. Shell and BP senior executives, unencumbered by antitrust restrictions, met privately and at the British Foreign Office to share information and develop strategies. Shell was known for the best market intelligence, whereas BP was considered more closed but also more successful at finding oil. The American majors, on the other hand, had to find subtler ways to communicate. During the 1950s, facing repeated charges of collusion from the Justice Department, they had grown increasingly wary of direct contact. By 1959 they were so cautious that a high-level British attempt failed to produce a unified industry strategy in preparation for the Cairo congress. Shell's John Loudon, however, was not too disappointed. He told BP and the Foreign Office in March 1959 that during a recent trip he had been "blunt" with U.S. executives about not making any price move before discussing it with Shell and BP. At a follow-up meeting in May, Shell and BP reported their pleasure with the outcome of the Cairo congress: there would be no retaliation for the February price cut. Loudon and BP's Chairman Bridgeman still hoped their U.S. counterparts might agree on a common strategy to avoid being "played off against each other." They would soon find out, however, that their American rivals did not feel obliged to notify them. Jersey would soon take action that the British would come to rue.[34]

London, nevertheless, remained at the heart of the international oil world. Wanda always stopped there on her travels eastward. Despite the loss of empire, the city had recovered from the wartime bombings and immediate postwar deprivations. Although the Aramco partners had eclipsed their British rivals as preeminent players in the Middle East by the mid-1950s, London was the city of choice for intelligence gathering and industrial espionage, the key listening post for the majors' Middle East experts and headquarters for some of their joint ventures such as IPC in Iraq and the Kuwait Oil Co. For the American oil elite, it was also something of a safe haven from the Justice Department. At home, they no longer felt comfortable even joining rival executives for a game of golf, but in London's discreet, well-appointed restaurants, such as the Mirabelle and the Connaught Grill, they could meet more freely, probe for information, and parse the latest oil news. The European glamorization of big business added to London's appeal as a city of robust romanticism. A British television series, *The Troubleshooters*, featured an international oil company called Mogul, managed by a group of sleek, heroic executives who outwitted rivals and averted calamities with cunning and class. The women were seductive, the Americans naïve, the diplomats pompous and ignorant, and the sheikhs unscrupulous.[35]

The Middle East destinations for international oilmen were also exotic: Beirut, Tehran, and Cairo. Beirut's nightlife—the cuisine, the casinos, the pickpockets—combined the chic of Paris and the Riviera with an Arabian Nights setting. Egypt's pyramids, teeming bazaars, and Old World charm enchanted Europeans and Americans alike, whereas Tehran attracted the more adventurous. Travel was still arduous throughout the region, and communications were limited to telegrams and the rare telephone call. Visa restrictions and travel bans on certain nationalities complicated matters, as did the occasional riot or outbreak of full-scale warfare: in Cairo in 1956 and 1957, Baghdad and Beirut in 1958, then Jordan and Yemen.[36]

American Middle East experts such as Jersey's Howard Page, Aramco's Terry Duce, Mobil's Brandon Grove, and Socal's George Ballou, all friends of Wanda's, were part of this worldly group of oil troubleshooters. Cold war uncertainties added to their status, as ambassadors and intelligence officers sought them for their in-depth knowledge. They "felt they had a responsibility, a patriotic duty, to tell the government as much as they could," Grove's son, himself an ambassador, recalled. Sophisticated and learned, these men also knew how to tap into a group of American diplomats called Arabists. Often the children of missionaries, these diplomats had grown up in the Middle East, spoke Arabic, and were attuned to cultural and political nuances. As they probed for details, the oilmen would filter out the idealism of the Arabists, many of whom sympathized with Middle East nationalism.[37]

Washington insiders were also willing to trade influence for information. The CIA was particularly interested in profiles of key people: Who were they? What made them tick? Who was influencing the shah and Nasser and, for that matter, Tariki? "Politics was key to the international game of oil," Brandon Grove Jr. said. "And it was an exciting game to be in." Though they worked for competing companies, these oilmen shared an esprit de corps. "It was a remarkably open society within a very tight circle," recalled the CIA's James Critchfield.[38] Yet to varying degrees, a certain postcolonial mentality prevailed among both British and American elites. Only they could truly understand the big picture, the cold war threat, the intricacies of the international oil trade. Still, despite this hauteur and their skepticism about Wanda's growing sympathies for the oil countries, they could not afford to ignore her.

A SINGULAR PLAYER

Wanda wrote her own script in this drama. "She was of that world but not part of it—at least not within the structured relationships—discreet, not conspiratorial," Critchfield said. Her access to the captains of the oil industry was unprecedented. She now met regularly with Shell's leaders, John Loudon, Sir David Barran, and Sir Frank McFadzean; BP's leaders, Sir Maurice Bridgeman, Sir Eric Drake, Sir Peter Walters, and Lord Strathalmond (Billy Fraser to her); and leading American executives Rawleigh Warner, Al Nickerson, William "Tav" Tavoulareas, Jack Rathbone, and George Parkhurst—all presidents, chairmen, or directors of the major oil companies. She scribbled these names over and over in her appointment diaries. No business journalist had ever cultivated this kind of long-term, in-depth access to the key decision-makers in any international industry.[39]

In the late 1950s, business journalism remained quite undeveloped in the United States and also in Britain. Although the *Economist* was well regarded and the *Financial Times* was improving, business journalists had none of the prestige and influence of political reporters. Industrialists ignored them; diplomats avoided them. "The less you poke your nose into our business, the better we will be" was the industry's prevailing attitude, according to Shell's David Barran, who made an exception with Wanda when he met her in New York in 1958.[40]

In Britain as in the United States, Wanda was the only reporter regularly allowed into the "inner chambers of the demi-gods," said Shell's John Bishop. Even the *Economist* senior editor Jack Hartshorn conceded that he never got the access Wanda had. She alone "won the confidence of the most difficult, most egotistical, most prickly people in the world," said *New York Times* energy reporter William Smith. Loudon, Shell's chairman, said, "Wanda was the only journalist I talked with and met regularly—the only reporter who really knew the business." Nearly forty years later, he still had a vivid image of her—tiny, energetic, the "sort of person you just don't forget." When he traveled to New York, he made a point of taking her to dinner or the ballet. She rarely wrote anything down, relying instead on her "near-perfect" memory and ability to compartmentalize what she learned. A meeting with her was a genuine exchange of information. BP's chairman, Eric Drake, recalled that "she would tell me sort of off the record what the others were saying. And then, no doubt, she would go to someone else and say what I was saying. She was quite open with me about it. In fact, we used to pull her leg about it." She kept "this insider role of hers very quiet." Eventually Drake became "rather switched off" by articles that, he said, gave "undue attention" to the oil nationalists.[41]

By penetrating the boardrooms in London, Wanda was, in effect, crossing class lines as well as breaking gender barriers, since business journalists were not then considered members of the upper class. "The fact that she was a woman may have helped her," said Lord Christopher Tugendhat, a former member of Parliament and now university president who started his career in 1961 as a *Financial Times* reporter. "They found her harder to type-cast. They weren't accustomed to journalists approaching them on the basis of social equality the way she did." Although there were a few women writing for the *Economist*, Wanda was in another league. At that time, a professional woman could not be both serious and glamorous, said Tugendhat, but she was both.[42]

Wanda made the most of a genuine turning point in the power politics of oil. "The dramatis personae were not very well known to each other," said Tugendhat. "The game hadn't got established rules or established moves. Everything was still fluid." Because of the oil club's isolation, Wanda's information and analysis, whether fresh from the Middle East or from private company dining rooms, was important, "the stuff of spy novels, but real life, too."[43] In effect, Wanda allowed herself to be something of a conduit, an interlocutor—a classic journalistic technique but one that was potentially toxic. "Sowing seeds in journalists' minds that might sprout up when they are talking with your opposite number," said Tugendhat, could be construed as "using" a journalist. And reporters might be unaware of the value or underlying implications of the information they received. But for them, information was currency. The best journalists gained more than they gave. It was a symbiotic relationship, this information exchange — "a high-flying game," said Youssef Ibrahim, energy reporter for the *Wall Street Journal* and *New York Times*. He and others, such as Anthony Parisi, followed in Wanda's footsteps, "but Wanda did this on a very big scale, bigger than any other reporter on oil or OPEC, before or after."[44]

She was both feared and sought out. "Everybody will tell you they never, *never* disclosed anything secret to Wanda," explained Alfred Munk, a senior Amoco executive, but the questions they asked her gave away a lot. "Wanda could often smell things out without their being said," Munk recalled. "She had a very cagey relationship with her sources." Those who talked with her regularly had to contend with suspicions. Not only were her good friends Brandon Grove and George Ballou accused of leaking to her, but even John Loudon had to fend off criticism from other Shell directors. In Aramco, there were actually "big arguments" about talking with her, recalled Joseph Johnston, a senior vice president who tried to limit Wanda's access because she put "too much information out there."[45]

Wanda's fiercest critic was Augustus ("Gus") Long, the granite-faced chairman of Texaco. The most tight-lipped and tight-fisted of the majors, "Texaco had a horror of the press," recalled Henry Moses, Mobil's representative on Aramco's secret executive committee, the Excom. Moses admitted that he was "brainwashed" by his colleagues, including Jersey's Page and Socal's Parkhurst, not to reveal too much to Wanda. Texaco's Excom member, Harvey Cash, was the most adamant. Because of Mobil's reputation of being "as close to Wanda as any company," Cash tried to intimidate Moses into staying away from "that Wanda."[46]

For her part, Wanda generally avoided Texaco, but the company occasionally drew her ire. Simply trying to get basic information on publicly known company operations was "harder than trying to interview Khrushchev," she charged in a scathing 1959 critique of an unnamed oil company. "If they don't like what I write about them in the future, they better not blame me." Only at the last minute was she strong-armed by *Petroleum Week*'s editor, LeRoy Menzing, into changing the Texaco name to "Company X." She made the criticism more general with the headline, "This Shoe Will Fit Several Feet," but most everyone knew who "Company X" was.[47] It was a prelude to a more serious showdown with Texaco a few years later, when Gus Long would seek his revenge.

GAINING CREDIBILITY

If some oil executives were less than charmed by Wanda, they grew even more disturbed by her series of reports on the leading oil nationalists after the Cairo congress. She had gradually come to question the "sanctity of contracts" argument the majors made over and over—that the concession agreements as legal documents could not be changed. During the 1956 Suez crisis, she had essentially agreed with that notion, but by 1959, she believed the companies ought to adjust those contracts to respond to the nationalists' concerns. For the companies, this shift in her thinking proved more than unsettling. For some, it was almost treasonous.

By 1959, Aramco's frustration with Tariki had also reached a new high. Since the 1940s, the company's well-developed public relations staff and its Arabian Affairs Division, a secretive intelligence-gathering group, had devoted significant amounts of money and energy to promoting a positive image of the company, even paying reporters and writers to contribute to this effort. But the Saudi oil director came to haunt them. He became, as one historian put it, Aramco's "public enemy number one."[48]

Annoyed by the press coverage that Tariki drew on a U.S. trip in May 1959, Aramco management blamed Wanda—that was the word from the CIA's Kim Roosevelt and Aminoil executive Harley Stevens, who said that Aramco believed "it is all Wanda Jablonski's fault for giving Tariki such a build-up." Stevens himself insisted that she was not to blame; Tariki would have gotten coverage anyway. But Aramco also complained about the arrogance of BP oil executives Hubbard and Chisholm for alienating Tariki in Cairo with their attempt to "talk sense" to him. These men, the Aramco executives said, by starting off with the "when you have been in the oil business as long as I have, my boy" line, "had already done more harm than good."[49]

Wanda enjoyed the notoriety. Together, she and Tariki stole the show—at least for one evening in May 1959—at the fifth World Petroleum Congress, a weeklong trade show in New York. At BP's opening-night cocktails at the Essex House Hotel on Central Park South, Tariki arrived after the party was in full swing, a radiant Wanda on his arm. The couple made a dramatic entrance to the murmur of the crowd as they walked down several steps into the main reception hall. He wore an elegant suit, she a dark, sleeveless cocktail dress with straps crossing at her neck. Tariki began to greet people "with Wanda still literally on his arm," recalled her colleague Chick Squire. Tariki did not let go until he reached an Arab journalist he knew, who in turn presented him in Arabic to Squire, who had reported for Beirut's *Daily Star* before joining McGraw-Hill's *Platt's Oilgram News*. Delighted to meet someone who had lived in Beirut, "Tariki threw his arms around me," Squire said, "dislodging Wanda in the process." Squire worried that Wanda, whom he regarded with some awe, might take offense. She did not.[50]

But the scene is significant as further evidence of Wanda's attachment to Tariki. She was already something of a legend within the industry, so just knowing her personally "was a real cachet," said Shell's Silvan Robinson. By taking Tariki's arm, she was making a statement to the assembled oilmen: He was not the gadfly that so many in their club considered him to be. His ideas had merit. Simply by socializing with the oil nationalists—the "ragheads," as they were sometimes called—she helped give them respectability. Some oilmen, said U.S. ambassador Richard Nolte, even "considered her a traitor" for her friendship with Tariki.[51] No longer struggling to be taken seriously, Wanda enjoyed the fact that her very presence gave her power.

Rumors were rampant, then and later, that Wanda was having an affair with Tariki. Not only did she talk openly about staying at his home in Jeddah, but she had been seen after midnight with him in bars in Beirut and Cairo, and he had visited her New York apartment several times. Yamani remembered wondering about them: "They became good friends, extremely good friends. Some people told me that they became intimate." But Yamani knew Wanda's style. In later years, he, too, would stay up late with her at her apartment and he, in turn, invited her

Wanda and Tariki at cocktail party, World Petroleum Congress, New York, May 1959.

to stay at his homes in the Saudi mountain resort of Taif or in Sardinia, where she vacationed for days, if not weeks at a time. She befriended his family. "That's what you do when you become good friends," he said.[52]

Other friends, particularly Feisal Mazidi, the young Kuwaiti official, and Alirio Parra, a former Venezuelan oil minister, doubted that she had an affair with Tariki. "If she had been willing to have affairs, she would have had one with me," the debonair Mazidi said. It was something she did not do. "Wanda and I were so close," Parra recalled, "but she kept some things separate from others. Between Mazidi and myself, we can vouch for that. I remember several nights, especially one in Beirut, when we were out late for dinner and then drinks. We talked and talked and drank, but I can tell you, nothing happened." Some oil-executive friends agreed. They did not see her having affairs with contacts. "She was too smart for that," said Amoco's Munk. "It would have been the kiss of death in her profession." Years later, Wanda told Beverly Lutz, Parra's ex-wife, that she did not have affairs with sources because it would have given them power over her.[53]

The time she spent with Tariki during the New York oil congress gave her attention but also some good material on his joint plan with Perez Alfonso. Their prospects in mid-1959, however, did not look good. It was hard to tell companies to rein in excess production when those countries wanted more cash and greater market share. The oil nationalists had four "culprits": Kuwait, where the BP–Gulf Oil joint venture kept setting new export records; Qatar, whose ruler made clear he wanted more output; Iraq, whose new regime was publicizing IPC's plan to double its production over several years; and Iran, where the shah, she learned, was insisting that oil output grow faster than anywhere else. What's more, North Africa was about to unleash a flood of oil, with stunning repercussions. By 1965, Saharan production might exceed one million barrels a day. The glut would continue to grow.[54]

MORE THAN A GENTLEMEN'S AGREEMENT

One man, at least, was willing to do with less: Perez Alfonso told Wanda that Venezuela would accept a "leveling off" of production. During several meetings in the garden of his modest Caracas home in September

1959, she questioned and debated this passionate "philosopher" of oil. "He is not and has never been a businessman," she explained in a column. "He tends to disapprove of money." A slight man with thinning hair and weary eyes, he rarely talked as openly as he did to Wanda, said Alirio Parra, his onetime aide. Wanda was special; she had remembered him during his exile. His country, he told her, needed to expand its industrial base and to rely less on oil and conserve it for future generations. He wanted Venezuela to set up its own national oil company, run on a commercial basis without special privileges, unlike those in Mexico and Brazil. By starting small, Venezuelans would "learn how to manage the business, not just draw rent from it," Perez Alfonso said. "It is easier for our people to learn from their own experience than from the foreigner." He then told her he was trying to persuade the Middle Eastern countries to join him in "an international compact" to restrict output and stabilize prices. That was significant, said Parra, because "these words—international compact—did not appear in the Gentlemen's Agreement."[55]

Perez Alfonso knew the majors would hear his message through *Petroleum Week*. Oil executives and diplomats in London and New York might not be willing to meet with him, but they did read Wanda's articles—as did the shah. Thanks in part to her coverage, Perez Alfonso's plan for a Texas-style "prorationing" for world oil attracted the attention of the Texas independents, who were still struggling against the majors despite the import quotas. Invited to speak at their May 1960 conference, Tariki and Perez Alfonso explained their plan in greater detail. But with Wanda, Tariki was even more explicit. Their plan to seek production restraints was not just about oversupply. It was also about power. The major companies "do not like us to do anything," he told her. The oil countries "have never had the right of saying to whom they are going to sell and how much. They have no authority." Wanda generally sympathized with his arguments, but she would soon dispute his position on government "prorationing" to reduce the oil glut. She was too much of a market capitalist for that.[56]

ANOTHER PRICE CUT

Despite the nationalists' protests, the majors still reigned over the oil world. Still first among the majors was Standard Oil of New Jersey, whose chairman, Jack Rathbone, had little patience for the oil nationalists. Rathbone had sought out Wanda's views, listened to her blunt assessments, and read her reports from Cairo and Caracas, but he had a business to run. What mattered to him was what was happening in the market, not the views of Abdullah Tariki or Juan Pablo Perez Alfonso. He would soon seek out Wanda's help—but with a different opponent.

As winter turned to spring in 1960, the glut got worse. With the independents underselling their way into established markets, the majors had to give larger and larger discounts on the official posted price for Middle Eastern crude to hold their ground. As a result, Jersey sold more barrels, but profits dropped. "It was so damned competitive that we were at our wits end almost to see how we could lower our costs another half a cent a barrel so we could make a little bit more return," Howard Page later testified before Congress. The large profit margins of the 1950s had indeed shrunk, though not as badly as the companies alleged. By July, Wanda was reporting on increasingly fierce marketing ploys, including promises to build refineries in return for contracts. From what she was hearing out of Spain, Thailand, and India, the chase for new deals "keeps getting dizzier and dizzier."[57]

Jersey had to do something to stop the carnage, Rathbone concluded. For the majors to deal more effectively with these new competitors, posted prices would have to come down enough to minimize the effect of these discounts. But he had a problem, one he considered more significant than the likely outburst from oil-producing countries: he was faced with opposition from within his own company. He knew that some members of his board, all Jersey executives, would vigorously oppose him.[58]

Unlike his quiet and conciliatory predecessor, Rathbone was opinionated and decisive. Promoted to chairman in May 1960, he brought both a change of policy and of style. Known for his "phlegmatic determination," as *Fortune* put it, Rathbone sought advice but then made up his mind "on the facts, stripped of ribbon and lace." He was a chemical engineer who had worked for Jersey for four decades and had turned the refining process from "a combination of guesswork and art" into a science. "With the husky frame of a roustabout and the craggy features of an amiable pugilist," Rathbone presided over a company that, according to *Fortune*, had become "a kind of Roman Empire" of the modern business world.

But the *Fortune* article elided a weakness that worried some of Rathbone's colleagues and competitors. Inexperienced with international issues, the new chief executive was tone-deaf to the clamor of the oil nationalists and insensitive to the longer-term implications for Jersey's vast foreign reserves. All that really mattered to him was Jersey's market share and profit margin. He saw no need to worry about the budgetary concerns of foreign governments. "Money," he said, "is heady wine for some of these poor countries, and some of these poor people." Since they had benefited enormously from the rapid growth in oil production, it was only fair that they should bear some of the burden of the decline in prices.[59]

As chairman, Rathbone began looking for ways to bring the board around to his point of view. His leading opponent was Howard Page, Jersey's chief Middle East expert. Though not opposed in principle to lower posted prices, Page adamantly objected to a unilateral cut just then. A small, soft-spoken Californian, this chemical engineer had international expertise from his years in Europe and the Middle East, where he helped put together the Iran oil consortium in 1954. A deft negotiator, Page was known for keeping a slide rule in his lap "so that he could calculate down to the last half cent on a barrel," wrote Denis Wright, soon named Britain's ambassador to Iran. "But he was also a man of some vision, and was very well able to understand other people's vision." Although Wanda frequently sparred with Page, she admired his sophistication and intellect. She had also known him socially since the early 1950s; they had been dinner guests at each other's homes.[60]

What bothered Page about Rathbone's position was the producer governments' possible reactions to another price cut. The uproar in 1959 over the BP-led reduction could not be ignored. On several occasions, Page had already asked the Jersey board to consider some revision of their Middle East agreements, "some sort of easing" in anticipation of the day when the market was so oversupplied that posted prices would have to be reduced. The board "told me not to be silly, the governments hadn't asked for it," he later remembered. "Never give them anything they don't ask for." By 1960, Page staunchly opposed Rathbone's plan for a unilateral price cut.[61]

Rathbone, however, decided to take an unusual step. In mid-July, he asked to meet with Wanda privately, without anyone knowing—especially his colleagues. He would give her an extraordinary scoop, but she had to protect him—no one could know he was her source. She could cite him as a top executive; that was all. She evidently agreed. But what about opposition from members of his board? "Howard Page, the know-it-all," Rathbone told her. "I'll show him."[62]

To ensure Rathbone's anonymity, Wanda apparently made a round of phone calls to her leading sources to probe about the possibility of a price cut and hint at what she was hearing. Knowing Page's previous problems with the Jersey board, she asked to meet him personally at his Rockefeller Center office. Page was an important source, and she wanted him to stay that way. He read the draft of Wanda's column, she later told Daniel Yergin, "in stony silence." Then he looked up at her with a cold stare and asked, "What ass gave you this?" Page knew she would not have written it without a clear signal from the highest levels in the industry. Wanda, of course, was inscrutable.[63]

Rathbone seems to have convinced Wanda that another price cut was coming, and her follow-up with other sources underscored the likelihood that his view would prevail. Indeed, Page's reaction did not dissuade her. What's more, her report on the significance of Rathbone's views had prompted action once before: her article on Jersey's 1953 policy reversal on oil prices had triggered a price cut—and a congressional investigation. As Jersey chairman, Rathbone technically had a stronger hand to play now than he had seven years earlier, but because of Page's opposition, he chose to work through Wanda.

She initially tempered her explosive message with caveats. In the first of two columns published in late July, she warned that the world oil price structure was under assault. Although it would be "foolhardy" to predict an imminent cut in posted prices, pressures had reached "a dangerous point." The latest evidence of oversupply problems came from

India, where the Soviet Union was making a bid to break into that market with cheap crude. The majors had to respond with discounts. "As of mid-week," she reported, "at least one major supplier had offered to reduce" its price by 27 cents a barrel to an affiliate in India. (In the spring of 1960, posted prices ranged from $1.50 to $2.31 a barrel, depending on quality.)

Despite big discounts from the independents, the system of selling crude to affiliates at posted prices had held together until then—but no longer. "We realize that this decision on India represents a very basic policy change," an unnamed top executive told her. "In effect, we had to cross the bridge on the question of discounts to affiliates—for we realize this may spread." Moreover, he added, "The tremendous amount of price discounting to third parties is bad enough. If it now spreads to affiliates, the fat's in the fire."[64] The message was unmistakable: U.S. prices were about to be cut.

By the following week, the fire had spread. With more evidence in hand, Wanda published a longer column that spelled out why a price reduction was imminent. She found a "growing conviction in top oil circles" that allowing "phony" posted prices to continue in the face of massive discounting was "demoralizing." Heightened uncertainty created a vicious downward spiral of discounting. "The more chaotic the market gets, the more likely you are to undershoot it," said an unnamed "key" executive. The greater the uncertainty, the deeper the discounting. It meant that companies had "to resort to the most involved kind of commercial intelligence to find clues to what your competitors all over the world are up to—and you're still never really certain."

Oil-producing governments were certain to protest. No one "is unmindful of the political repercussions of a price reduction in both the Middle East and Venezuela," Wanda wrote, although increased production would cushion the blow. Markets were "deteriorating badly." The glut was vast and long-term. Pressure from Russia's increasing exports was not the root of the problem, though it might be the "final straw" that could "topple the whole Middle East price structure." Summing things up, she again gave the final word to a "top" executive: "I don't see how the present fictitious postings can endure much longer. It's only a question of when."[65] Although Rathbone remained anonymous, his voice resonated in both columns. She quoted this unnamed executive at far greater length than she normally would one source, except for such authoritative figures as Nasser or Perez Alfonso.

She raised only one important criticism of the executive's argument. If prices did come down, would that have any substantial effect? Wanda had her doubts. Given the enormity of the surplus, "there's little hope that the present dog-eat-dog fight for outlets will disappear in the foreseeable future." As for the oil-producing countries, she acknowledged their budgetary plight but chided them for insisting that price cuts were designed just to take revenue away from them. They, too, had to learn the hard lessons of the market. Ever the realist, she seemed to accept the inevitability of the price cut.[66]

Tariki and Perez Alfonso were outraged. The oil-producing countries' revenue share was determined by the posted price, not the actual discounted price at which the majors transferred crude to their affiliates or the market price for third-party companies. John Pearson, *Petroleum Week*'s Caracas correspondent, specifically remembered Perez Alfonso's reaction when he was summoned for an explanation: "What was Wanda doing?" Did she not realize she was acting as a mouthpiece for the majors by preparing the intellectual and political climate for a price cut?[67]

Tariki, dismayed and hurt, told her so directly. "Dear Wanda," he wrote from Jeddah. "I think you knew more than your predictions and I am really afraid that the cut is coming because these are the signs for it." Why couldn't the oil companies curtail production and thus strengthen prices? "Why should the M.E. [Middle East] oil [be] posted as if all of it is going to N.Y. while only [a] very small amount of it is going there?" The solution was nationalization of Saudi Arabia's oil industry, which Tariki euphemistically called "integration." He was also upset with her criticism of the oil countries: "Thank you for the compliment or is it an insult? About us in the producing countries being ignorant or unrealistic. You put it in a more refined way as usual." He wished she would "stay friendly with the producing nations but, I see it is very costly to be so. I sure admire, though not like, your writings recently." But he was convinced that one day she would agree with him.[68]

Oil executives and diplomats from London to Beirut were also on the alert. An American embassy official in Lebanon called Onnic

Marashian to find out how Wanda had gotten wind of such news. The official was "really upset" because a price cut would fan the flames of Arab nationalism. Shell and BP executives were angry, too—especially at the lack of notice or consultation. Despite John Loudon's careful rounds the year before, they received no warning other than Wanda's columns. Unaware of Howard Page's opposition, BP's chairman, Wanda learned, "bawled out" Page for Jersey's unilateral decision.[69]

On August 9, 1960, without any notification or consultation with the oil countries, Jersey announced cuts of about 7 percent in posted prices on Middle East crude to $1.46 to $2.13 a barrel. It was a "fatal move," Loudon later said. "You can't just be guided by market forces in an industry so essential to various governments." But Rathbone had evidently succeeded with the Jersey board. Although reports about what happened at the final meeting are contradictory, the outcome is not. Some Jersey directors said the board had backed Page's motion to trim posted prices only after some sort of agreement had been reached with the producer governments, but Rathbone overruled it. Others insisted the Page motion was defeated. In either case, Rathbone managed to impose his will.[70]

In the process, though, he had resorted to a tactical move that would have been anathema to John D. Rockefeller: He played the press card. Concerned about Page's opposition, Rathbone took the risk of publicizing his arguments through Wanda's columns in order to consolidate the board behind him. By anonymously setting out his views in print, he apparently hoped to create extra pressure for a decision in his favor— and it worked. Although Wanda did not openly support his position, her extensive quotes from this "top" executive overshadowed her one substantive criticism (on the long-term effectiveness of the move) and presented his arguments as a viable response to difficult market conditions. At the very least, she focused the debate; she pressed the oil world to address the issue.

That Rathbone would resort to such a significant leak was evidence of a developing new relationship between big business and the press. In this instance, as in others over the years, Wanda's reportage did not change the course of events, but it did likely accelerate the pace of developments.

FORCING THE SHAH'S HAND

Jersey's move did not go unanswered. Despite the dismay expressed by British and Dutch oil barons, all the major companies followed with their own reductions in posted prices of Middle Eastern oil, though the reductions were somewhat smaller than Jersey's. Within two weeks, prices had settled at about the level they had been in 1950. In Caracas, Jeddah, Baghdad, Tehran, and Kuwait City, however, the news led to a different kind of response: a flurry of telegrams, hastily arranged meetings in various capitals, and finally a conference scheduled for mid-September in Baghdad to develop some kind of collective Middle Eastern response.[71]

It was by no means certain that all five countries—Saudi Arabia, Venezuela, Iran, Iraq, and Kuwait—would actually consent to meet. Iran proved the most hesitant. Some directors of the NIOC, the Iranian state oil company, had privately denounced Farmanfarmaian for signing the Gentlemen's Agreement in Cairo the year before. Still bitter about the lack of Arab support during Iran's aborted nationalization effort, they preferred to maintain their distance. The shah, for his part, was uncertain about his power and his policies. He did not know whether he should send a delegation to Baghdad, as some NIOC directors had urged.[72]

Once again, Wanda found herself in a position to influence a policy decision, or at least its timing. In late August, shortly after the Jersey price cut and before the Baghdad meeting, an encounter between Wanda and NIOC Chairman Entezam in New York revealed just how unsure the shah was. Her detailed notes on the meeting also give insight into the way Wanda coaxed her sources into revealing more information than perhaps planned and how, in some instances, she would tailor her private comments to fit her audience.[73]

Entezam was an impressive source, the kind of person, she believed, with whom the companies should deal forthrightly. A handsome Persian diplomat posted to Washington in the 1920s, he was once married to a free-spirited American from a well-known Georgetown family. Entezam had been minister of foreign affairs when he first met Wanda in Tehran in 1953. As the NIOC's leader in 1960, he was so highly regarded that the shah did not usually interfere. Known for his "quirkiness, brilliance, and absolute integrity," according to Denis Wright, this chain-smoking intellectual was also a gifted mathematician, an expert in optics, and a dervish—a mystic who disdained materialism. Wanda liked him.[74]

The NIOC chairman was quite realistic about the commercial reasons for the price cut, Wanda wrote in detailed background notes about their initial, off-the-record meeting. What mattered most, Entezam said, was how much the majors were disproportionately cutting production in Iran more than in Iraq, when Iran had three times the population. Wanda had heard that line many times, so she pushed him: "Look, you've got something else in mind. Why don't you be frank with me?" Smiling, he nodded and replied that the companies must realize that Arabs were totally unreliable. They should be careful not to continue treating Iran as an Arab state, which Iranians deeply resented. She, in turn, could speak authoritatively about what the companies were planning without naming names. She had "every reason to believe," she told him confidentially, that Iran's production would rise in the future, not fall.

To pin him down further, she pushed again. "You and I are lucky in one respect," she said. "I'm not an oil company, so you and I are not negotiating with each other. We're just chatting informally. With me you don't have to worry about what you say, or how you say it, since none of it is official. Tell me, in specific, concrete terms, just what it is you really want in the way of production." He replied with three points: Kuwait's production should not grow at all in 1961; at least one half the total increase in Middle Eastern crude that year should go to Iran; and under no conditions should Iraq's output exceed Iran's—a reference to a longstanding rivalry between the two nations.

Since their meeting was off the record, she then tried to persuade him to make these points publicly. His commercial sense was refreshing, she said, in contrast with some "asinine" comments that Perez Alfonso and Tariki had recently made about the need to freeze prices and allocate production worldwide. Restraining the market, she believed, would hurt all sides. It was true that most Americans did not know the difference between Iranians and Arabs, but publicizing his views would help clarify the difference. Cajoling him with one argument after another, she made her pitch with seeming nonchalance. "Entezam was silent for a long time," she recalled. But she had played him well. "Maybe you have a point there," he finally said. After consulting with Tehran, he agreed to an official interview in which she pushed him even further: Would Iran, she asked, become actively involved in sharing information among the oil-producing nations, as stipulated in the still-secret Gentlemen's Agreement? Entezam flatly said no.[75]

Entezam then cabled the NIOC, saying that "Wanda was writing a big article in *Petroleum Week* that Iran would not cooperate with the other producers," Farmanfarmaian recalled. Entezam had given Wanda what had been the official Iranian position: Tehran wanted the majors to consult on prices, but most of all, it wanted higher—not lower—production. But much had happened since Entezam had left for London and New York. Iranians and Arabs alike were protesting the arbitrary price cut of August 9. The shah himself, in a speech on August 19 —the seventh anniversary of the U.S.-initiated coup that overthrew Mossadegh—denounced the companies for their unilateral decision. Tariki and Perez Alfonso were pressing him to send delegates to Baghdad for a meeting of oil-producing countries that wanted to challenge the majors. Though the shah resented the Arabs for taking advantage of Iran's losses during nationalization, the majors' unilateral price cut made him far angrier. A placatory visit from Jersey's Howard Page and BP's Harold Snow only further upset him. They "made no serious effort," Wanda later learned, "to talk things over with the NIOC." And some NIOC directors, including Fuad Rouhani and Farmanfarmaian, argued that if Iran participated in planning some kind of oil-producer organization from the start, it might have more control over the group's decisions and have extra leverage to use against the companies.[76]

Confronted with Entezam's cable about Wanda's forthcoming article, "the shah suddenly changed his mind," Farmanfarmaian recalled. He wanted an Iranian delegation to attend the Baghdad gathering. The shah sent Farmanfarmaian as his personal representative, while the NIOC board sent Rouhani and Fathollah Naficy. In New York, an embarrassed Entezam asked Wanda to kill the article.[77]

What came out of Baghdad would forever change the relationship between the international companies and the nations where they extracted oil. On September 14, 1960, after five days of negotiations, Saudi Arabia, Venezuela, Iraq, Iran, and Kuwait announced that they had formed the Organization of Petroleum Exporting Countries (OPEC) to challenge the unfettered power of the international oil companies. Never again, they proclaimed, would the majors dare to lower the price of crude oil without consulting them. Few Western reporters, however, were allowed to cover the event, as Iraq's internal political problems had put the regime on high alert. Marashian tried to get an entry permit but was denied, and Wanda was not invited. OPEC's leading cofounders, Tariki and Perez Alfonso, were upset with the "free market" attitude of her columns, which, they believed, enabled the majors' second price cut. But although not present at OPEC's creation, Wanda had been intimately involved in its creation, which earned for her the tongue-in-cheek title "OPEC's midwife."[78]

7.

Smoking Out
the Shah

When Wanda returned to Tehran in November 1960, some Iranians still thought she was a CIA agent. She could not go anywhere or meet anyone without being noticed: dining with oil officials such as Naficy and Rouhani, drinking whiskey with Farmanfarmaian, or even just going to the hairdresser, she was followed. The agent who reported her every move to Iran's intelligence service was Rostam ("Mike") Bayandor, a National Iranian Oil Co. "guide" and interpreter. "Everybody suspected everybody in Tehran," Bayandor later explained. She did not make his job easy; her late hours exhausted him, and she drew a lot of attention. When the NIOC gave a reception in her honor, several newspapers published photos of "this recognized authority on international oil affairs." In a dark, tailored dress with a geometric collar set off by a strand of large pearls, a glamorous Wanda was surrounded, of course, by a host of men.[1]

None of this pacified her; she was there to interview the shah, and he was making her wait. "The day after tomorrow," she was repeatedly assured. This went on for weeks until late November, when she scribbled a note to her colleagues on the tear sheet of an Iranian newspaper: "Can't write much for the moment as wrist sprained & in sling. Expect interview with Shah tomorrow." The interview indeed did occur, and it proved to be one of the best of Wanda's life—and so significant that it changed the oil world's agenda. It was a serious setback for OPEC's co-founders and, instead, reassured the leaders of the oil club.[2]

SHOCK TACTICS IN BEIRUT

Before traveling to Iran that November to interview the shah, Wanda had already begun to assess the new organization called OPEC. Although the U.S. and British press gave it only cursory coverage, OPEC's creation in Baghdad was "history-making," she wrote in an otherwise noncommittal commentary from New York. So far, the founders had not produced a plan of action, but they had publicly denounced the price cuts and committed themselves to consulting one another. Privately, however, Wanda was furious about the price cut, Onnic Marashian remembered. The majors had rational reasons to reduce the price, she believed, but the cut was badly handled. Jersey, especially, should have shown the producer governments some respect and talked with them first.[3]

Wanda set off in late September for the Middle East, taking her usual route: London, Beirut, and then Tehran. Over lunches and dinners in London with leading oil executives, she heard over and over how foolish Jersey was to lower posted prices without notice—although, of course, BP had done just that in 1959—and how irrational the nationalists were to think that OPEC could impose production constraints.[4]

When she arrived in Beirut, site of the second Arab Petroleum Congress, in October, she found the mood electric, the debate vitriolic. The fireworks missing in Cairo a year earlier were on full display in Lebanon's capital in 1960. For Tariki and Perez Alfonso, the timing of the second congress was as fortuitous as the first. In the aftermath of the second price cut and the founding of OPEC, the oil nationalists had a formal platform from which they eagerly attacked the majors as imperialists. When company officials declined to respond, Wanda noted privately, the Arabs became convinced they had "the big companies scared."[5]

Tariki gave what Wanda called a "riveting" speech in Beirut in which he presented a more explicit plan to restrain output than the one he had discussed privately with her a year earlier. He wanted to divide markets on a hemispheric basis and force the oil companies to divest themselves of their international subsidiaries. This plan sounded like "price dictation and complete managerial control of the oil industry," Wanda wrote in a commentary. It was virtually a blueprint for a cartel, although the oil-producing countries did not have the power to enforce their wishes.[6] When she tried to get an interview with Perez Alfonso, he would not talk with her; he and Tariki were still upset over her recent *Petroleum Week* columns. From their standpoint, she not only helped bring on the August price cut and gave only the industry view, but Perez Alfonso had even denounced her reports publicly, saying they were part of a "press campaign" to promote lower prices. She had, in fact, reported their objections and their official response to Jersey's decision. Although still angry with her, Tariki gave Wanda an exclusive interview after his keynote speech because he wanted her to pass on an additional message to the majors: if they did not comply with his demands, the majors would have only five years left in the Middle East.[7]

Wanda now felt that Perez Alfonso and Tariki had gone too far. Although she had done much to promote their views, had helped them hone their arguments, and had bluntly told the majors they should deal with the oil nationalists as equals, not inferiors, she was too much of a free-market adherent to countenance these new demands from her old friends. She explained their latest plan in some detail in a *Petroleum Week* column, then knocked down each point with practical counterarguments. How would Middle East governments give up access to Western Hemisphere markets? How could the majors be expected to accept such curtailments? What if production cuts led to much higher prices, driving consumers back to coal? Tariki made "no provisions for such market losses," she noted, nor had he explained how the majors would agree to give up the most profitable part of their business—the producing sector—through divestment. His new definitions of "international proration and price stabilization," she concluded, were "a far cry from what these terms mean when used by the Texas Railroad Commission."

The chasm between the oil nationalists and the Western oil companies had widened. Thanks to Jersey's unilateral decision to lower posted prices, the dialogue and moderate tone of the Cairo congress had

turned into demands and threats. Indeed, the primary focus had become OPEC's willingness to challenge the majors, rather than broader
concerns about redistributing oil wealth.[8]

Not all oil nationalists, however, had become so militant. One of
the up-and-comers, the young Kuwaiti Feisal Mazidi, spent long hours
talking with Wanda. Years later, he recalled being worn out by Wanda's
persistent questioning and her penchant for late-night scotch: "She
wouldn't quit." In Beirut, they would rendezvous at well-known bars
such as the Café du Roi or Madame Baroine's in the Commodore Hotel. "She was not openly pro-Arab," he said, "but she was close to us."
Because of her reputation—"because she was so very precise in getting
the truth"—she persuaded him to go on the record with his more moderate views.

Mazidi, Kuwait's first graduate from a British university, had both
benefited and suffered from studying in the West. A slight young man
with thick, dark-framed glasses and a cleft chin, Mazidi grew up in a
mud hut. Late-night reading by kerosene light ruined his eyes, he said,
but he excelled in school, which led to a scholarship in England. A spe-
cially arranged meeting in London with Billy Fraser, then director of
the BP–Gulf Oil joint venture, the Kuwait Oil Co., led to a haughty
dismissal after only a few minutes. Fraser was totally uninterested in
him, Mazidi recalled. When he then tried to get a job with Kuwait Oil
in Kuwait, manager Jimmy Scott told him there was "no need to get
Kuwaitis involved in running the company." Mazidi found work in
Kuwait's Department of Finance in the late 1950s, and soon became assistant oil director, but, like Tariki and Parviz Mina, he never forgot the
humiliation.[9]

Only a few years later, when Mazidi was preparing to attend the 1960
Beirut oil congress as one of Kuwait's two delegates, Scott hurried over
to warn him not to meet with Wanda. "You should never allow yourself
to be quoted by that *woman*," he said. Mazidi, with the ministry's approval, delighted in doing exactly that. In an exclusive interview, Mazidi
told Wanda that Kuwait was concerned about the latest proposal from
Tariki and Perez Alfonso. Although his government strongly endorsed
OPEC as a way to develop a unified stance, Kuwait did not want a production cut.[10]

Wanda personally disagreed with the Tariki–Perez Alfonso plan, but the conflict it aroused made for potent material. As usual, she was much sought after for lunches, drinks, and dinners, and as she made the rounds, Wanda readily argued with all sides. Brandon Grove, who also attended the Beirut congress, noted in his diary that she had debated not only one Arab delegate until he felt "pretty indignant," but had given Grove himself, her dear friend, "a terrific lacing" on the "real and fancied shortcomings" of his arguments. Wanda could be tough on everyone.[11]

FALLING FOR A MARRIED MAN

To add to the tumult of those heady days in Beirut, Wanda apparently fell in love. The object of her affections was Jack Hartshorn, a slender, soft-spoken Englishman who looked a bit like her father. A witty intellectual and a senior editor at the *Economist* with a good prospect of being named editor in chief, Hartshorn made the Middle East one of his specialties after his wartime service in North Africa and Egypt. He later went on to write two well-respected books on the oil industry and also became a consultant. But in 1960, although the *Economist*'s Beirut "stringer" reporter, Kim Philby (whose work as a double agent for the Soviet Union would soon be exposed), was reasonably good, Hartshorn wanted to cover the Arab oil congress himself.

He had met Wanda in Beirut the year before. "I liked her immediately," Hartshorn later remembered. "She looked like she was out of a fashion magazine—always immaculate, even if she had just gotten off the plane." They saw each other almost daily for two weeks in Beirut, according to Wanda's appointment diaries. At Brandon Grove's invitation, they also joined him for a sightseeing expedition to Damascus, a two-hour drive from Beirut, where they visited mosques, lunched at the Orient Club as guests of Grove's friend Mohamed Ahdab, and indulged Wanda, who wanted to shop for Damascus brocade and other "junk" in the markets.

During that time, her New York office got worried. Tariki's electrifying speech, Mazidi's interview, and a few commentaries came in from her during the Beirut congress, but then silence. For several weeks, she sent nothing—no copy, no letters. When a cable arrived from LeRoy Menzing, *Petroleum Week*'s editor, asking for some sign of life, Wanda quickly wrote an apology, explaining rather lamely that she had taken two weeks' vacation time. It was not worth trying to see oil people in Kuwait or Libya after the Beirut congress because "everybody who was anybody would be busy entertaining the 'brass.'"[12]

A married man, Hartshorn later declined comment on whether his relationship with Wanda went beyond friendship, but he did not deny it. Marashian distinctly confirmed it, having seen the pair together in Beirut a number of times that October, holding hands and showing considerable affection. "It was a sort of puppy love," Marashian said. "It was so unlike Wanda. I remember being so surprised to see her that way. Here is this woman who goes into palaces and sees kings and rulers, and she's falling for this guy?" Four decades later, Marashian recalled that "the only affection—not friendship or closeness—but the only real affection I ever saw Wanda show was for Jack Hartshorn." Since Wanda kept no personal diaries, wrote few personal letters, and rarely spoke about her feelings, little is known about her relationship with Jack Hartshorn. Later in life, though, she confided to a friend that he was the one "true love" of her life.[13]

AFTER THE CIGARETTE, ASKING FOR MORE

Given the militant tone of the second Arab oil congress in Beirut, it was non-Arab Iran, Wanda realized, that had become the most critical player in OPEC's immediate future. And given the success of Iran's more militant oil officials in persuading the shah to send a delegation to Baghdad in September 1960 to participate in the creation of OPEC, Wanda needed to be in Tehran to probe for a better understanding of just what potential OPEC might have in challenging the oil club. Besides, the shah, through a letter from Rouhani, said he would grant her an audience. (Rouhani, who had gone to the opera and several dinners with Wanda in New York earlier that year, had asked, in a scribbled note on the letter, Was the charming lady at the House of Vienna "still singing about the woman being smarter than the man?")[14]

At first, her prospects for copy were slim. Even though she was Entezam's guest, "none of the government officials," she wrote her editor, "are ready to talk for publication until I first see the Shah and *they*, as well as I, find out what the 'policy' is." She could not get much out of company sources either. Executives from one subsidiary, she complained, would "run the other way" when they saw her. She dutifully went on another inspection tour of the Abadan refinery and new production centers, but they did not merit major coverage. Entezam himself was not ready to chance another statement. He insisted on waiting until he learned the shah's position on "price cuts, prorationing, OPEC, etc.," and he asked Wanda to push the shah on these points. The NIOC still had limited influence over Iran's oil development, since most of the industry—from exploration and production to refining and exporting—was under the majors' control through the consortium.[15]

When Wanda finally got her interview with the shah, it did not begin well. Wanda, according to her "shadow," Bayandor, did not like the shah, and the monarch was suspicious of her because she was a woman. Twenty minutes, she was told, was all she had. At first, the shah—not yet the absolute ruler he would become—launched into a lengthy explanation of what his country needed. Because Iran was using its oil revenue wisely, and because of its large population, it deserved a faster production rate. For Wanda, this was nothing new.

After a few minutes, she stopped taking notes, she later told friends. She stared at the chain-smoking monarch and began to tap her pencil, but he droned on. No longer able to restrain herself, she changed the subject by asking for a cigarette. His attendants, she explained, had confiscated her pack, saying, "You can't smoke in front of the Shahinshah [king of kings]." Of course she could smoke, the shah replied. A pack of cigarettes was within easy reach, but she just looked at it. Startled, the shah got up from his ornate chair to offer her one and returned to his seat. Then she pointed out that she did not have a match, either. Once again, the shah "forced himself to stand and lit my cigarette." As she would later say only half-jokingly to her friends, "*This* is how you have to deal with monarchs."[16]

Cigarette now in hand, Wanda asked the shah if she might make an observation. "Yes, of course," he replied, according to her detailed notes. If she were Iranian, she might resent Kuwait's high production

Wanda meeting the shah of Iran before her long-awaited interview, November 1960.

rate, or even Saudi Arabia's, but have some sympathy for Iraq because its oil growth rate was even lower than Iran's, "and this, without even having gone through a nationalization shutdown."[17]

The shah "grinned, as if he enjoyed the joke." Actually, Iran's rate of growth, she continued, had been substantially higher over the last five years than Iraq's. The shah leaned forward. "Are you sure?" "Yes, Sir," she replied, "quite sure." His advisors had the statistics. The shah responded by spelling out what he wanted from the consortium: to assure Iran one-half of the total growth rate combined for all Middle East producers "at the very, very least" and to restore Iran to its former position as the region's leading oil producer. This, Wanda knew, was unrealistic, but she decided not to contradict him—at least not yet.[18]

Suddenly, the door to the royal chamber opened. An official announced that the prime minister needed to see the shah right away "on a matter of great urgency." The shah, "in a peremptory tone," dismissed him, saying he was busy; the prime minister could wait. The interview continued for an hour more.

The companies' recent price cut, the shah said, greatly angered him. Neither he nor the NIOC had been consulted in advance. "Remember, we have not accepted the last price cut." (Wanda "let this point pass by," she wrote in her notes. "The top NIOC director had told me just the opposite, and I thought best not to help freeze the Shah's position on this point.") But with some prodding on her part, the shah said that although he understood the oil consortium's commercial concerns, it had to start treating Iran as an equal partner. As owner of the oil, Iran had a right to know what revenues it could count on for its national budget, but the consortium would not divulge its financial records or forecasts. He and his advisors "should not be left with the feeling that everything is being hidden from us."[19]

Wanda quoted him directly on this subject, but the article made no mention of several other times during the interview when she corrected him, though these are chronicled in her notes. When the shah said that if the oil companies had "sound" reasons to cut prices in the future, "all the governments will agree," she asked whether he would mind if she disagreed. The shah laughed, saying, "Not at all, but why do you disagree?" Iran might have very knowledgeable oil officials, she said flat-

teringly, but some Arabs under Nasser's influence were much less experienced in oil economics and would say "no" for political reasons. "The
Shah," she noted, "sat silent for a long time, looking thoughtful, but said
nothing."[20] Wanda smoked and stared.

To deflect her response and her gaze, he spoke effusively of the "real
partnership" Iran had with two independent oil companies—Standard
Oil of Indiana's subsidiary Pan American and Italy's state-owned Eni.
Both had agreed to give Iran 75 percent of their profits, instead of the
usual 50 percent. Here, too, Wanda had a problem. "This seemed like
a minor showdown that had to be faced," she wrote in her notes. She
knew the shah was oversimplifying the case; since these two companies
had yet to find oil, it was premature to compare their contracts with the
consortium. She explained the technical reasons, and then took a jab at
Iran's "so-called Italian partner." Eni had just agreed to import more
Russian crude at such a steep discount that it would undermine Iran's
prices, while, in return, Russia was getting steel pipe to help it export
even more oil, she said. Was that a good partner? Again, silence.

Again, the shah broached a new subject. He wanted the consortium
companies to make the NIOC a partner in some of their European refineries. Would Iran, she asked, be willing to pay its share of the refineries' capital costs? Wouldn't the shah prefer to focus his investments
instead on Iran? Another long silence. Finally, exasperated with her, he
said, "Well, maybe this is too full of difficulties and too complex," but
the consortium had to give Iran more than 50 percent in profits. Was
this an official demand? "No," he answered. "It would raise too many
difficulties." Suddenly worried by what she had led him to say, the shah
backed off. The entire discussion of the profit split, he then insisted, was
off the record. But that did not stop her. She went on to correct him
on other issues, even suggesting at one point that he check with the
NIOC's legal department.[21]

The interview, said Bayandor, took on the form of a tutorial, "with
Wanda explaining the views of the oil companies and the Shah asking
the questions." Sir Denis Wright, Britain's ambassador to Iran, concurred: "The Shah was still willing to learn then. It was later that he became utterly impossible." Wanda was also surreptitiously explaining key
issues, including the profit split and Iran's compensation obligations, at

the NIOC chairman's request. Entezam was "anxious that she should use her reputation in the oil world to put the Shah straight on these matters, since Entezam did not feel that he himself could do it," the British embassy's economic counselor wrote in a cable to London. Iran's most senior oil official, Entezam, evidently believed this Western female journalist could influence the shah more effectively than anyone else.[22]

Wanda then brought the shah back around to the most critical subject for OPEC: whether he supported the new version of the Tariki–Perez Alfonso plan to restrict production. She first got the shah to talk again about why Iran needed more oil, and then she deliberately provoked him. If increased output was so important, she asked, why had his government agreed to "restrict production and set export quotas," as the Venezuelan oil minister had announced? The shah immediately said that Iran had not agreed to any such thing. When Perez Alfonso made that statement, "we denied it," he replied. Cautiously, she corrected him, saying his government had "not actually issued a denial"—just an indirect notice that left the issue open to interpretation. In London and New York, Perez Alfonso was quoting "the Shah himself as having agreed on export quotas."

The shah was appalled. The Venezuelan's statements were "absolutely untrue." Was that official? "Yes," he replied. What about Perez Alfonso's suggestions that he "had reached a 'back door' secret agreement with the OPEC countries" not to cut into each other's "natural markets," but instead divided them up between Eastern and Western Hemispheres, thus keeping Middle Eastern oil from being shipped to the United States at prices that would undercut Venezuela's exports? "This is also untrue," the shah answered. "What nonsense. The only thing we agreed to was the official Baghdad resolution, which was published." Was this statement also official? "Yes," he replied firmly.[23]

That was it. She had her lead.

"The Shah of Iran this week dismissed international oil proration as 'impractical,'" Wanda wrote in *Petroleum Week* for December 9, 1960, "although he acknowledged the usefulness of the newly formed Organization of Petroleum Exporting Countries." For the oil world, the implications of this story were significant. Less than three months after OPEC's creation, the shah of Iran had undercut the immediate threat this ambitious organization posed to the international oil system. How-ever angry he was with the majors for their lack of consultation, the shah wanted them to make Iran the Middle East's top producer again, not cut output, restrict market share, or set quotas. OPEC could be useful for sharing information and developing joint policies, but not for con-fronting the companies with rationing through an international Texas Railroad Commission. With this interview, Wanda again forced the shah's hand, although this time more deliberately than she had with her August interview with Entezam. She also gave him a tutorial on the finer points of market conditions and contract terms. In return, she got cigarettes and a scoop.

UNSETTLING EVERYONE

The interview with the shah made quite a splash; major papers from London to New York to Caracas picked it up.[24] But Wanda found her-self in a delicate position. Despite the accolades, she was also openly criticized, even threatened. As the bearer of an unfavorable message, she had further alienated her nationalist friends Tariki and Perez Alfonso.

The following week, she also upset the majors. She criticized the consortium companies for their high-handed ways, for their continued unwillingness to bring Iranians into a real partnership—a point she had been hammering since 1954. Among Iranians, there was "growing an-noyance—almost bitterness—over NIOC's lack of participation," she noted in one column. Consortium decisions remained "so shrouded in secrecy and mystery" that they caused "excessive suspicion and distrust." Far too many Western expatriates were still running Iran's oil opera-tions, with far too little training of Iranians. The consortium, she in-sisted, simply had to do more.[25]

Blunt as she was in print, she was even more so in private. If the consortium did nothing to remedy the situation, "there might be a very serious break ahead," she warned the British embassy's economic coun-selor before she left Tehran. The majors had to make NIOC a partner by taking "it fully into confidence." The delay in handing over nonbasic operations to Iranians, plus the "apparently complete lack of consultation at the highest level," was "dangerous." Wanda was particularly struck by the fierceness of the cultivated, well-educated Rouhani. At one lunch with her, he pushed his plate away in disgust, saying he could not understand why the Western companies did not learn lessons from the past. Ten years earlier, he had warned Anglo-Iranian officials about the likelihood of nationalization six months before it happened but was "laughed off." Why would the majors still not listen to them? When she returned to London and New York, she promised Rouhani, she would give the consortium's executives her frank assessment. As she then told the British diplomat, if key NIOC officials continued to feel "resentment at being treated like children by the Consortium," they might try to pull out of the 1954 settlement, and the whole structure would collapse.[26]

Senior British officials had mixed reactions. In cable traffic between the Tehran embassy and the Foreign Office, they voiced concerns about the ramifications of Wanda's interview. Despite the shah's explicit statements on price and production controls, her column would "give fresh impetus to OPEC" simply by giving it so much attention. They worried that her private account to the economic counselor of the shah's off-the-record views showed "a potentially very dangerous strain in consortium/ Iran relations"—much worse than a BP executive had recently told them. Though the shah's interview was "depressing" since he "trotted out all the old hares," one diplomat said in an urgent cable, the "firm and outright way in which Miss Jablonski dealt with him" was gratifying. The shah's categorical rejection of price and production controls "cannot fail to have considerable effect on the other members of OPEC." It confirmed that Iran was the "weak link in OPEC."[27]

Tariki and Perez Alfonso, of course, were even more upset with Wanda. Her interview with the shah had shattered their hopes for quick, unified action against the majors. OPEC as a simple consultative body would have no clout. The timing of Wanda's article was also awkward for Perez Alfonso, who was about to host OPEC's second meeting only a month later, in January 1961. Quickly trying some damage control, he told the Caracas newspaper *La Esfera*, "I don't know why, but she has turned into a great opponent of OPEC." With the headline "Miss Wanda Jablonski Harpoons OPEC," the paper published a personal attack, quoting the minister as saying that "his friendship with Miss Jablonski was unchanged" in spite of her "sharp spears against OPEC." Her report from Tehran distorted the shah's views, Perez Alfonso told *La Esfera*. His insistence on higher production for Iran did not mean he was against "prorationing." What's more, Wanda presented "the idea of price stabilization as a crime of which Perez Alfonso is responsible." Charging her with "calculated maliciousness," *La Esfera* claimed she even became "Wagnerian" when she quoted a company official saying that if Perez Alfonso's plan were carried out, the majors would have only one year of operations left in the Middle East. For that, *La Esfera* issued her a threat: when Wanda arrived in Caracas for an OPEC conference in mid-January 1961, she would be greeted "with the finger-pointing slogan: 'People know her.'"[28]

DEFINING THE DEBATE

Wanda, it turned out, had no plans to attend the conference. *Petroleum Week* had already assigned its Caracas correspondent, John Pearson, to cover this second OPEC meeting. Instead, once she had pinned down the shah with her interview, she went on to visit other key oil-producing countries to gauge the significance of OPEC. She found that there was serious skepticism about OPEC in Kuwait, a founding member, and in Libya, a prospective one. For Perez Alfonso and Tariki, her regular cables to New York with breaking news and commentary meant more bad news.

But Wanda had uncomfortable news, too, for the companies. In a remarkable coup, according to the *Economist*, the "redoubtable Miss Jablonski" had extracted some highly sensitive information out of Kuwait—extensive details of a precedent-setting agreement between Kuwait and Shell for a new offshore exploration concession. She not only explained what was significant about the new deal and why, but even got Kuwaiti officials to reveal enough about each of the nine competing bids that she could present them, side by side, in a chart.

The companies were shocked, as were journalists, diplomats, and government oil officials. Never before had a Middle Eastern state revealed so much about its deliberations. In London, Hartshorn wrote to

Wanda, Shell had "played it awfully close to the table"; despite considerable effort, he had not been able to get any details. In Kuwait, Ashraf Lutfi, a Palestinian in the ruler's secretariat, was equally surprised. How did she do it? "A masterpiece of detective work," he wrote effusively. "Of course I did not like it that so much information has slipped into your hands and has been published, particularly details of the bids by other companies." Kuwait's London representative also praised her: "There must be something about you which makes strong, silent men disclose what ought to be kept secret and I am not surprised that *The Economist* should call you 'the redoubtable Miss Jablonski,' although I should have chosen a term less formidable and perhaps more agreeable to feminine ears."[29]

That description had, in fact, concerned her. "What did you mean by that?" she later asked Hartshorn. He replied, "I mean you are very good, very direct—by comparison with the British."[30]

Kuwait's decision to pick Shell, she explained in her December 1960 exposé, heralded a new, more sophisticated relationship in the oil business based on "greater realism" on both sides. For the first time, one of the majors had agreed to a joint venture with a producing country with terms more commercial and more reflective of a true partnership than previous ones were. Although Shell's financial terms were slightly less attractive than some competing bids, other factors had tipped the scales. "Without any doubt," she wrote, "it was Shell's huge world-wide marketing facilities—and know-how—that constituted its chief 'sex appeal.'"[31] Leaders of this British protectorate, which would not become independent until 1961, expressed strong interest in developing Kuwaiti expertise so they could stop "being just landlords and tax collectors." Sheikh Jaber al-Sabah, the minister of finance and senior member of the royal family who would become ruler in 1977, talked with Wanda extensively, though he remained an anonymous source. The new, partially state-owned Kuwait National Petroleum Co. (KNPC), he explained, would start with domestic distribution and then, with Shell's help, expand overseas.[32]

Off the record, Wanda gained considerable insight into Jaber's future plans. In a book published in 1962, Hartshorn wrote about an interview in December 1960 between Jaber and an unnamed female

American reporter "whose knowledge of oil concessions in the Middle East is prodigious." It was Wanda, of course, he later acknowledged: "Who else could take on a sheikh like that?"[33]

In the interview, according to Hartshorn, Jaber casually mentioned that the KNPC might eventually begin its own exploration activities. "But where?" the female journalist probed. Perhaps in onshore fields relinquished by Kuwait Oil, the BP-Gulf group, he replied. "But Your Excellency," she said, "I was under the impression that the Kuwait Oil Company's concession did not make any provision for the relinquishment of unexplored territories, as for example those in Iran and Saudi Arabia do." Jaber smiled. "You are right," he answered. "It does not. But I believe you attended the oil congress in Beirut." During one of the sessions there, Billy Fraser, Kuwait Oil's director and a lawyer by training, had explained that when contract problems arose with Kuwait, "We simply talk it over as friends, and settle it as friends." Only after that, he said, did they "call in the lawyers and bring out the agreement." So Jaber asked her, "Did you not hear Mr. Fraser say that when we have something to discuss as friends, we are not concerned with what is or is not in the agreement?"[34]

Once again, Wanda prompted yet another "strong, silent" man to reveal more than perhaps he had planned. Within less than two years that retort had become fact, when Kuwait pressured the BP-Gulf group into relinquishing fully half of its acreage in 1962. Wanda had zeroed in on a nuance that, with further prodding, led to the revelation of an important policy change. Although she could not publish the news that December and she had to protect the source, she used it well in her information-trading game.[35]

She did manage to publish another scoop, however, that sparked controversy—Kuwait's "second thoughts," as she put it, about the Tariki–Perez Alfonso "prorationing" plan. Ahmed Sayid Omar, chief of Jaber's oil department, said the main reason OPEC was founded was to find a way to stabilize prices. The more he considered the plan to ration output, the less he thought it could work.[36] Sayid Omar was not alone; this more moderate tone also came from the oil nationalist Ashraf Lutfi, who had originally endorsed the production-restriction plan. The more Lutfi studied the issue, he told Wanda, the more skeptical he became.[37]

Although Kuwait was then the Middle East's biggest oil exporter, the protectorate was little known. By quoting its leaders at length, Wanda presented them to her Western audience as thoughtful and insightful. Winning the confidence of Jaber, Sayid Omar, Lutfi, and Feisal Mazidi was an "extraordinary feat," Mazidi later recalled. Jaber at that time was "very socialist in his thinking." Sayid Omar, he said, was a more old-fashioned, highly cultured man—"an *ancien régime* type" who often recited poetry. Mazidi and Lutfi, on the other hand, were ardent oil nationalists who admired Tariki. Wanda persuaded them all to talk with her. "By writing down our views respectfully," said Mazidi, "she was espousing people like us."[38]

Wanda then spent the week before Christmas in Beirut. She saw oil contacts and friends for lunches, drinks, and dinner every day, and went to the hairdresser, but she stayed in her hotel Christmas Day. It was easier to deal with holidays, she later told friends, by traveling.[39]

On the day after Christmas, Wanda flew to Libya, which had a remarkable story of its own. In just four years, Western companies had discovered eight prolific fields there—an exploration pace "unequaled anywhere in the world." Although completion of its first major pipeline to Europe that year would bring a surge of crude-oil exports, exacerbating the worldwide glut, Libya had "no intention of joining OPEC for the foreseeable future," she learned in an interview with Mohammed Asseifat, the chairman of Libya's Petroleum Commission. To make matters worse for the oil nationalists, North Africa's most promising petroleum exporter had no interest in production restrictions. It wanted just the opposite—as much foreign exploration and investment as possible.[40] Relations with the companies, Libya's de facto oil minister said, were "very, very good." Asseifat sympathized with OPEC countries seeking better terms, but he also understood the commercial pressures on the companies. "I'm sure the president of Jersey Standard didn't cut his prices last summer just for the fun of it," he told Wanda. Libya "believes in the sanctity of contracts."[41]

With this slew of articles and commentaries over four months from Beirut, Tehran, Kuwait City, and Tripoli, Wanda undermined the hopes of OPEC's founders, at least for a time. As Hartshorn put it in a January 1961 essay for the *Economist*, "statements made in exclusive interviews secured by Miss Wanda Jablonski" marked a retreat from the more hostile tone at the Beirut oil congress. These changes in attitude "could become significant" as OPEC's leaders tried "breathing substance into this paper organization" at their meeting in Caracas in January.[42]

Privately, some Arab moderates were pleased. "You have blown up OPEC single-handed," Egyptian lawyer Saba Habachy wrote her approvingly. "You have brought confusion to the ranks." Though sympathetic to nationalist concerns, Habachy, whose clients included Western oil companies, opposed the "impracticable dream of an oil industry directed and controlled" by a government cartel. As for the offensive contracts, he preferred negotiations, not ultimatums.[43]

Rather than destroy OPEC with these interviews, however, Wanda actually helped put it on the West's strategic map, at least the oil map. As Britain's diplomats in Tehran noted, her attention to OPEC gave it a certain prestige, a "fresh impetus."[44] Nevertheless, Tariki and Perez Alfonso blamed Wanda for putting their agenda in doubt, and the Caracas meeting in January 1961 only gave them further complications. After a week of debate, the meeting closed without a commitment to production restraints—or anything even close. The Persian Gulf state of Qatar, thanks to its burgeoning production, was readily admitted as a new member of OPEC, but controversies swirled over just about every other issue. As Perez Alfonso finally conceded, his production plan "may be inapplicable at present."[45]

Political tensions within their own countries further complicated the effectiveness of OPEC's principal founders. After the initial flush of success in Baghdad and Beirut, Perez Alfonso was stymied by criticism from Venezuela's business community, which "erupted both in Congress and in the press," John Pearson reported. As for Tariki, a new twist in the Saudi power struggle between King Saud and Crown Prince Faisal gave him added stature—and added uncertainty. King Saud, who wrested control back from his brother in December 1960, elevated the finance ministry's petroleum directorate to an independent ministry and named Tariki the country's first official oil minister. But neither Tariki nor Saud lasted long in these roles.[46]

In January, during OPEC's meeting in Caracas, Wanda finally

headed home after nearly five months on the road. On the way, she stopped again in London for several days; British and Dutch chief executives had sent telegrams to Beirut and Libya asking her to be sure to come and see them.[47] She also saw Jack Hartshorn for the first time since their days together in Beirut three months earlier. In a letter Hartshorn wrote to Wanda in mid-December, delivered to her Beirut hotel, he gave no hint of anything more than friendship. "Looking forward with love," he signed off. Anticipating Wanda's stopover in London in January, he also noted that his wife, Judy, was "anxious to meet you." Whatever had happened between the two of them in Beirut was over. Hartshorn met her at the airport, and her appointment diary indicates one lunch at the *Economist* and one dinner with him — but they were not alone on either occasion. It was Brandon Grove, not Hartshorn, who walked Wanda back to her hotel after lunch, and Grove who accompanied her to the airport. If Hartshorn was indeed the love of her life, it was, except perhaps for that brief interlude in Beirut, an unrequited love.[48]

QUESTIONABLE CONSULTING

Upon her return to New York in January 1961, bad news awaited Wanda in McGraw-Hill's Green Pagoda. Despite its editorial success, *Petroleum Week* had been hit hard by an economic recession and the oil glut. Although paid circulation in 1960, its fifth year of publication, exceeded that of the *Oil & Gas Journal*, the magazine did not draw enough ad money. McGraw-Hill believed that *Petroleum Week* was competing for advertisers who would otherwise buy more pages in its other magazines, especially *Business Week*. The magazine's expensive overhead in New York was also a handicap, and its star reporter was costly — not so much her salary as her expenses for travel and for last-minute cables at "urgent" press rates.[49]

Despite her prizes and her international acclaim, Wanda felt overworked and underappreciated, she told Georgia Macris, and she worried about her financial future. All she had gotten from her divorce, she told friends, was a muskrat fur coat. Yet professionally and socially, she associated frequently with wealthy oilmen and "behaved like a CEO," as one said, though she did not have the income to match her lifestyle. She wanted and needed more money. Her solution was to get paid for what she was already doing informally—briefing senior oil executives on what she learned from her travels. In effect, she proposed to create a side job as private consultant while remaining a senior editor at *Petroleum Week*. This way she would be paid by check, not just by news tips.[50]

Surprisingly, management agreed. *Petroleum Week*'s publisher, Ervin DeGraff, in a January 1960 letter to McGraw-Hill's vice president, Harry Waddell, said he had discussed with Wanda the "dangers of entanglements with outside clients," but concluded that her journalistic integrity would not be compromised. Her briefings would be tightly circumscribed, and she would keep DeGraff and *Petroleum Week*'s editor, LeRoy Menzing, fully informed. According to the contract, she would discuss only general impressions, not company-specific or confidential information, and she would retain "complete independence" in gathering material and writing articles. She would not provide written reports, exclusive analysis, or advice on securities.[51]

This was a lucrative arrangement. Wanda's clients—at least six companies—paid her $500 to $1,500 per quarter for about eighteen months from early 1960 until mid-1961, and she kept it quiet. Some colleagues later expressed amazement that McGraw-Hill approved the deal—and in writing. Indeed, Wanda was on questionable ethical ground as a journalist and was putting her reputation at risk. Although she had given considerable advice and information to her sources for years, she crossed a line when she became a paid consultant, jeopardizing her independent status and her credibility as a journalist. It is not clear why she took this chance, but her lifelong worries about money and her anxiety about *Petroleum Week*'s prospects must have contributed to this serious lapse in judgment.[52]

Typewritten notes from that time give some insight into how Wanda briefed her paid clients. In August 1960, for instance, she learned from Mohammed Salman, director of the Arab League, that his president had asked him to become oil minister. He told her off the record that he might accept the position, but was worried about going back into politics and would miss his job in Cairo. Salman's wife, who joined them for lunch, kept saying, "They don't pay him anything, and anyway I want to go back to Baghdad where we belong." Since she had known Salman for years, Wanda believed he would likely accept, as he did several months later. Although she could not publish the off-the-record information, she may have used it to tip off her clients to this possibility.[53]

Wanda may have also briefed her clients in early 1961 on what she considered to be the oil companies' inept participation in the first two Arab oil congresses. Her recommendation, as spelled out in her private notes: Except for a few people from Middle Eastern operating companies such as Aramco, stay away. "The best solution, I believe, always is let the Arab handle the Arab." The moderates would likely keep the radicals in check: "They're not fools. It was [Kuwait's] Sheikh Jabir himself, as well as Sayid Omar, who individually and separately told me how they deplored how the Beirut meeting had been turned into a political diatribe. At that time, I replied that oilmen would be fools to come to another meeting like that and they both nodded agreement and understanding." She did not, however, oppose all public relations by the oil companies. To the contrary, "the industry must do something, and do it fast."[54] These views, apparently meant for a private company audience, dovetailed with her published commentaries but were nonetheless given privately for a fee.

DEMISE AND DEBUT

When McGraw-Hill announced in April 1961 that *Petroleum Week* would cease publishing in May, most of the staff was stunned, caught completely off-guard. Wanda, more aware of *Petroleum Week*'s financial situation, had talked about this possibility with Marashian and another journalist friend, Fuad Itayim, in Beirut the previous October, but the news of the magazine's demise still left her crushed. She was forty and divorced, with no children, no family money to rely on, no job, and still recovering from her brief relationship with Hartshorn in Beirut. Without her reporting job, her consulting work was also in jeopardy. What was she going to do with her life? How was she going to make a living? It was the most serious professional crisis she had ever faced. She must have talked it over with her parents, but apparently they did not lend much support. Eugene had retired in 1957 with a sufficient but not generous pension. He quickly returned to his first love, botany, at the New York Botanical Garden, which named him an honorary curator. To be close to the garden, he and Mary sold their New Jersey home and moved to Riverdale, just north of Manhattan in the Bronx.[55]

Although Wanda still had dinner once a month with her parents, her relationship with her mother remained prickly. Sister Mary Louise Sullivan, a nun at the nearby Cabrini High School and a friend of Mary's, remembered hearing about "a real strain" between them. Passages in Mary's twenty-page private account of her husband's career, written in the early 1960s, suggest that she was both proud of and envious of her daughter's professional success. Eugene "gained the affection and admiration of everyone who worked with him," Mary wrote. "His success went to his heart, whereas Wanda's went to her head." Whenever it was that Wanda read that essay, placed among the few private papers she saved, it must have cut to the quick. Despite her usually composed demeanor, colleagues recalled, Wanda was quite sensitive to criticism. In any event, she could not and would not turn to her parents for financial help.[56]

Letters of support and condolence poured in. Those from the Middle East were particularly heartfelt. Mohammed Salman, by then Iraq's oil minister, found the news of McGraw-Hill's decision "discomforting" since he relied so much on the magazine, but he said he was sure Wanda would find another venue. Amir Abbas Hoveyda, a future Iranian prime minister, said her magazine "always seemed to be the only one I ever got time to read," while his colleagues at the NIOC worried what they would do without her "wealth of information."[57]

She considered a number of options. Possible jobs with *Business Week* and the *Economist* turned out to be too restrictive. She wanted at the very least a signed byline, but neither magazine would oblige; it was against editorial policy. Hartshorn was especially concerned about her prospects. She was too much of a star, he wrote, for her to work for the *Economist*: "A paper that does not carry signed articles by regular staff people would not really be an adequate platform for the best-known, and entirely redoubtable, columnist in the oil industry."

Wanda had discussed the possibility of a subscription-only newsletter in a conversation with Marashian and Fuad Itayim the previous fall. One of Wanda's chief complaints about *Petroleum Week* was its slow delivery overseas; a weekly newsletter, shipped airmail, would be ideal. It was her idea, Marashian said, but it seemed too big a venture to take on alone. Hartshorn, too, encouraged her to consider starting a newsletter. The *Economist* publishing group might be interested, he hinted, since it was already experimenting with one. Wanda tried hard to persuade McGraw-Hill to restart *Petroleum Week* as a newsletter on international oil. She showed the company's brass the flood of mail from her supporters. It would cost much less to produce and would provide analysis not available in the daily *Platt's* reports. Newsletters were rare at the time, but she knew that many *Petroleum Week* subscribers would pay a lot for her kind of intelligence. McGraw-Hill, worried about the prospects of a long economic recession, was not interested.[58] Wanda was crestfallen.

When Hartshorn came to New York on a long-scheduled business trip in June, he lifted her spirits, but the *Economist* did not come through, either. Oppressed by the heat and the weight of indecision as spring turned to summer, Wanda fretted. Out of work for three months, she could not wait much longer. She had no source of income other than her limited consulting fees. Her efforts to find a buyer for *Petroleum Week* or another suitable platform for her columns had failed.[59]

She did not want to become a full-time consultant; she was not particularly fond of that kind of briefing. She wanted to chase the news. She wanted the adrenaline rush from tracking down inside information and nailing a story—getting it right and getting it into print.

The trouble was that very few women in her day were publishers. A couple of notable exceptions—newspaper publishers Helen Rogers Reid and Katharine Graham—inherited that mantle from their husbands, but no woman had ever launched an important industry publication. What bank would loan her the money to start it up? Who would be willing to take a chance and work for her—an untried, untested female publisher? Could she even sell enough up-front subscriptions to her industry friends to get it started? But what other alternative did she have? Ultimately she wanted to be her own boss, not at the beck and call of others, not a subordinate. She was too much of a prima donna, and by this point in her career she had worked too hard.[60]

When Wanda floated the idea of starting her own newsletter with prepaid subscriptions, her best industry contacts enthusiastically supported her. Leon Hess, founder and chairman of Amerada Hess, who had known Wanda for many years, offered to advance her a large sum for start-up costs. David Barran, then Shell's New York director, and a number of other leading executives said they would pay in advance for bulk subscriptions to distribute to their various offices at home and abroad.[61]

These offers were tempting, but Wanda was anxious about her lack of financial security. How would she manage? "I will never forget walking up and down the streets of mid-town Manhattan with Wanda, listening to her hash out how she might start her own publication," recalled Napier Collyns, one of her Shell friends. Wanda was bursting with nervous energy and smoking incessantly as they walked and walked. "She was so worried about what to do," Collyns said, "but so passionate about the need for her kind of reporting." He tried to reassure her. "My buddies at Jersey and Shell wanted her to re-create what they used to get from *Petroleum Week.*"[62] However much they sometimes cringed when Wanda's reporting caught their companies off-guard, they knew they needed her intelligence, her interviews, her analysis.

These appeals worked. Wanda began to recruit people who might join her. Her first good prospect was not a journalist, but a former CIA operative—one of Kim Roosevelt's "cowboys," Jim Eichelberger. He would help her gather intelligence, and she and others would write it up. In late August, she began an almost daily round of lunches and cocktails to interview potential reporters and editors. Notably, Georgia Macris, her prime candidate for editor, and Onnic Marashian, by then in McGraw-Hill's London office, both turned her down. The outlook was too risky for Marashian, who had a young family, and Macris knew too well how difficult it was to work with Wanda. So she turned to others, including a reporter from *Oilgram News* and a bright, entrepreneurial analyst from Jersey. She also got a boost from McGraw-Hill's management, which gave her *Petroleum Week*'s subscription list for free. Wanda then flew to London, where her friends at Shell readily agreed to buy bulk subscriptions, though BP did not. Less than enthusiastic about her investigative reporting, BP executives would take only a handful, but she was able to line up several hundred subscriptions, with Jersey, Mobil, and Shell as her biggest supporters.[63]

By September 1961, Wanda had decided to make her move. She covered an industry built on the notion of taking big risks, and now she would have to take her own. She accepted the offers of bulk subscriptions from several majors, paid in advance, but turned down Hess; she did not want to owe too much to one person. "Wanda really liked Leon Hess," said Margaret Clarke, the Kuwait Oil public relations official who had shared late-night drinks with Wanda in Kuwait in 1957, "but she didn't want to be anybody's poodle." Bulk orders from some of the majors and a few orders from her Middle East and Venezuelan contacts would have to suffice. Unhappy with the alternatives and anxious to find a way to make a living on her own terms, Wanda decided to take what she called "my terrible gamble": she founded *Petroleum Intelligence Weekly*. Almost immediately, it became known as the bible of the international oil world.[64]

8.

Publishing
the "Bible"

Wanda courted controversy—often positive, sometimes nearly toxic— as she sought as much attention as possible for her newsletter. Within months of *Petroleum Intelligence Weekly*'s launch in November 1961, Fuad Rouhani, Wanda's Iranian friend who became OPEC's first secretary general, got angry with her for revealing details of a secret study prepared for the new organization. He said she stole it off his desk. The majors got upset with her, too. Frustrated by her scoops and her OPEC coverage, Aramco's executive committee imposed a ban on speaking with her, and some majors forbade employees from talking with her reporters. Fortunately for Wanda, most of her sources, especially senior officials, ignored these orders.[1]

She welcomed the attention, but two accusations, both personal, were so serious they could have derailed her. One threat had a familiar ring: in 1963, she was suspected of being a spy. In some ways, this was not surprising. Wanda cultivated an aura of secrecy. In published articles and private comments, she often hinted that she knew more than she would say. Her badinage and enigmatic personality alternately attracted and baffled contacts who would try, always unsuccessfully, to coax her into revealing her sources. So even though *PIW* in its first two years gave OPEC more substantive coverage than anyone else, some

OPEC officials—particularly her old friend Rouhani—became convinced that Wanda was a paid agent for the major oil companies.

The other accusation came that same year when Gulf Oil officials falsely implied that a *PIW* reporter was having an affair with one of their employees in return for information. As a result, two major oil companies—Gulf and Texaco—canceled their subscriptions and tried to get the other majors to do the same. But because Wanda had established *PIW*'s credibility and authority so quickly, not only did her reputation survive these challenges but her newsletter benefited from the unexpected attention over the cancellations. It was not the kind of publicity she would have sought, but she knew how to take advantage of it.

GETTING INTO PRINT

At its outset in 1961, *Petroleum Intelligence Weekly* was far from a guaranteed success. Indeed, Wanda was so dismayed with its poor quality in the first two months that she later had the staff's copies destroyed. But she had plenty of money for start-up costs and salaries. Having personally lined up nearly five hundred prepaid subscriptions, mainly from Shell, Jersey (formally renamed Exxon in 1972), and Mobil, and charging $365 a year for most readers, she amassed a considerable sum up front. (Annual subscription rates for other business newsletters were generally much lower.) Wanda just said, "A dollar a day—that sounds reasonable to me!" When *Forbes* later questioned her high price, she replied, "If my information is good, they'll pay any price for it. If it's no good, they won't take it at any price." But then she had to live up to these high expectations, a constant source of worry.[2]

Wanda and her entire staff assembled on a Saturday to stuff airmail envelopes with the first issue, dated November 6, 1961. Only a few weeks earlier, they had moved into a small, dingy office in midtown Manhattan, with one glassed-in room for the publisher and a newsroom with secondhand furniture and well-used manual typewriters. The staff and the format changed considerably over the first few months. Jim Eichelberger, the Middle East expert who had recently retired from the CIA, quit after barely four weeks. Although helpful in searching for staff, leasing the office, and lining up a printer, he was not a good fit. Wanda also fired an experienced newspaperman because he had a drinking problem, but she soon replaced him with Ade Ponikvar, a "rewrite man" from McGraw-Hill's *Chemical Week*. The entire staff rotated for copy-editing duties on press night, taking the subway to the Chelsea warehouse district to proofread the pages that came off old printing presses —mammoth machines that clanged and steamed from the hot lead. Unhappy with *PIW*'s initial appearance, Wanda switched to onionskin "bible" paper—much more expensive, but pleasingly lightweight. Folded in three, *PIW* then fit perfectly in a man's jacket pocket. The size was crucial, she insisted. At oil gatherings, she would stick a crisp new edition of *PIW* in the top pockets of her friends' jackets, so that everyone could see it. The fine paper added to its cachet.[3]

Initially, Wanda's big coup was hiring Jack Hartshorn. Setting aside her personal feelings when she realized their brief romance would go nowhere, she asked for his help in launching her publication. Hartshorn arranged with the *Economist* to remain one of its senior editors while contributing signed articles to *PIW* as a part-time European editor. Wanda even gave him a small financial stake in the newsletter as an inducement. John Buckley, however, turned out to be her best long-term investment. A rising young executive at Jersey, Buckley had known Wanda since the late 1950s, when he gave her some news tips. Frustrated with Jersey's bureaucracy, Buckley was eager to take on a new challenge. "Wanda trained me well as a reporter," he later said. ("What's significant?" she would ask repeatedly. "What does it mean?") A quick study and able administrator who became managing editor within two years, Buckley knew how to weather Wanda's mercurial moods, and he provided an important buffer between her and the staff. Like his mentor, he eagerly went after scoops: "I never got a story from Exxon about Exxon, but I got plenty of stories from Mobil about Exxon and vice versa. We also got a lot from the independents about the majors." Wanda was especially pleased when they reported scoops on two companies in the same issue—each would be upset about one story and eager to read the other.[4]

MARATHON IN EUROPE

Going to press those first few months proved agonizing. The initial party atmosphere quickly dissolved as the staff struggled to fill eight pages every week. *PIW* was printed in the boldface/lightface style developed at McGraw-Hill, so that busy readers could skim the highlighted first sentences of each paragraph for the most important news. Distracted by staffing, training, and production issues, Wanda managed to get some exclusives, but not enough. She would have to go to Europe and the Middle East to hire extra staff and dig up more stories. The results were more than worth the trouble. After a week in London that spring of 1962, she went to Geneva to cover OPEC's fourth conference. Although she chafed at having to wait in hotel lobbies during its closed-door sessions, she eagerly passed out issues of *PIW* to key officials — Perez Alfonso and Francisco Parra of Venezuela; Ahmed Sayid Omar and Feisal Mazidi of Kuwait; and several Iranians including Reza Fallah, who had been the manager of the Abadan refinery when she visited in 1954, and Fuad Rouhani.[5]

Abdullah Tariki was not at the Geneva conference. His alliance with nationalist rebels within the royal family of Saudi Arabia had led to his abrupt dismissal in March 1962. When Wanda introduced herself to his replacement, Ahmed Zaki Yamani, and asked for a private interview, he suggested she attend the press conference he was about to give. She did not go to press conferences, she replied, and she walked out to the hotel lobby, sat down in a strategically located armchair, and stuck her nose in a women's magazine. Yamani, she knew, would have to walk right by her to get to the elevators. After his briefing, he did just that, but went directly up to his suite. According to her account, he then sent down an assistant to invite her to join him. Would he give her an interview? No, he was playing cards. Tell him I'm not interested, she said. A few minutes later, Yamani's aide returned to invite her for an interview, but Yamani did not say much at that time. Ian Seymour, a fellow reporter and sometime rival at the *Middle East Economic Survey*, remembered talking in the Geneva hotel lobby with Wanda and Yamani. The new Saudi oil minister "smiled a lot and we made polite conversation, but we didn't get anything out of him then." Wanda, Seymour added, "of course knew a lot more about the oil business in those days than Yamani."[6]

Though Yamani did not disclose any information she could use, what Wanda learned in hotel hallways and private tête-à-têtes gave her considerable material for *PIW*. Perez Alfonso also provided enough exclusive information for a hard-hitting page 1 story. Then, in a bylined commentary, Wanda wrote that considerable internal strife existed within OPEC. Given their failure to develop a coherent plan of action against the majors, some delegates felt that "on prices they will soon have to put up or shut up." They simply did not have the power to force the majors to raise prices. Yet she also highlighted the delegates' frustration with the West—the majors and their governments—for refusing to acknowledge OPEC's existence and legitimacy, let alone promote its direct participation in the industry. Although a senior official of Kuwait, when interviewed by Wanda in December 1960, had expressed optimism about a partnership with Shell that would supposedly develop Kuwaiti expertise in oil operations, the year-old partially state-owned oil company, the KNPC, still had no crude production of its own. The majors controlled it all. "We're like a thirsty man looking at rivers of water all around us," Sayid Omar, the KNPC's first chairman, told her. Wanda argued that the majors' foot-dragging made no sense.[7]

Her industry friends acknowledged her point but deflected it, at least for a while. "Increasingly as time went on," David Barran remembered, "Wanda criticized the industry for trying to maintain the old attitude of keeping a barrier against the producing companies." Some oilmen, particularly Loudon and Barran at Shell, were more willing to recognize the problem than others. Exxon's Howard Page, according to Barran, "was almost the last defender of the doctrine that 'Posted Prices are the business of the Companies,' long after most of us had accepted the reality which Wanda herself had long been preaching."[8]

After three weeks of nonstop meetings and coverage from London, Geneva, Zurich, and Genoa, Wanda retreated for the Easter weekend to a pristine Alpine lake. The Italian resort, she wrote her New York staff, "luckily is now ugly with rain and cold enough for me not to resent too much spending Easter at a typewriter." She then headed off for a monthlong trek through Milan, Rome, Madrid, and Lisbon with a seemingly nonstop agenda of meetings and lunches, cocktails and dinners—and stories to write.[9]

A TOUGH BOSS

Even at such a distance, Wanda was an exacting boss. Between meetings and waiting for airplanes, she scribbled regular missives to her staff on hotel stationery, dissecting the issues, praising and criticizing, and responding to a torrent of questions—what to do about telex charges, how to deal with the printer, when to allow vacations, which "stringers" to hire, whether to buy air conditioners. Over and over, Wanda harangued her staff to be more precise, to double-check with sources, to use stories from stringers only if they could be verified, to be stylistically accurate and consistent: "At our high price, we can't afford to beg forgiveness." While she was "on the firing line in Europe," her main reporters were to read each other's stories and be jointly responsible.[10]

Although later known as a martinet with her staff, Wanda successfully cajoled and encouraged her reporters during her early years as publisher. "The issues look good, very good," she wrote from Geneva. "The Rag is holding up very well and I want to congratulate all of you for the wonderful job you're all doing," she then wrote from Madrid. "If I just keep criticizing, please don't get too irritated," she wrote in June after an entire page of corrections. "Remember, I warned you [that] on trips there simply isn't time to write all kinds of things one would like to discuss—only the things that need to be communicated urgently." Impatient with mediocrity, however, she was "a Tartar when it came to style and accuracy," Buckley recalled. When a reader found a mistake, "she took it personally."[11]

Halfway into her five-month European tour, the frenetic pace began to take its toll. So many foreign cities, noisy hotels, strange food, and command performances from early morning until midnight, seven days a week. "I'm worn to a frazzle," she admitted after more than two months. "Am rushed as the very devil," she scribbled in one message. "God, I'm exhausted," she noted in another. Little problems—changing foreign currency, wiring money, cabling stories, and even making phone calls—often turned into major headaches. "My time," she complained, "is cut so sharply by the endless hours wasted in getting anything *done* in Europe." Not only was she trying to get exclusive stories, but she also had the duties and worries of a first-time publisher.

To give herself a break, she planned to take a few days off in Lisbon.

Instead, she got a big story on Portugal's success in finding and developing oil in Angola—and no time to relax. By the time she got to Paris in early June, Wanda conceded that she could not "bear the thought of one more oil interview or of hearing about any more oil problems." When Ponikvar, the "rewrite man," commiserated with her for getting "interview shakes and shudders," she was offended. "Not that," she shot back. "Simply haven't *time* to do everything and am exhausted going at a seven-day a week pace, month after month." But she could not yet return home. She was still "frantically looking" for top-quality reporters: a full-time European editor because she needed more than Hartshorn could produce, plus more stringers, even if she could not be sure which ones would be best. "How the hell can one tell whether a stringer will work out just by looking at his puss?" she wrote in exasperation to her staff.[12]

Finally, in Paris, she scored another coup: she persuaded Helen Avati, who was "by far the best oil reporter" at McGraw-Hill and was widely respected by oilmen in Europe, to join *PIW*. As Paris bureau chief, Avati would report on France and North Africa and coordinate with Hartshorn in Europe. Wanda had wooed her by taking her to lunch or cocktails or dinner every day for a week. "It was like a bad Kafka novel, my husband said, because I didn't know what to do, and Wanda kept pressing me." A big raise, considerable freedom, and a byline on special reports finally persuaded her. "You don't want to work for Wanda, do you?" asked one of her McGraw-Hill bosses, who offered to match the proposed *PIW* salary. "Everyone knows what she's like!" But Avati had made her choice.[13]

An American expatriate who started as an assistant in McGraw-Hill's Paris bureau in 1953, Avati had become a reporter for *Petroleum Week* in 1955. Like Wanda, she was clever at developing senior government and industry sources, eliciting information without formal interviews, and tipping off her sources when she could. Avati would work for Wanda longer than any other reporter—from 1962 until she resigned in 1988. Over time, the two women came to dislike each other. Avati resented Wanda's tight-fisted management and deliberate slights. And Wanda, acutely aware of how skilled Avati was in digging up some of *PIW*'s best scoops, grew jealous over time, both personally and professionally. But in 1962, when Avati accepted the offer, Wanda was elated: "This was the biggest goal of my 'mission' to Europe and I feel like shouting 'Eureka.'"[14]

CHESS GAMES

Wanda's 1962 European marathon produced an exposé on an OPEC study of the international oil industry that caused no small controversy. Although word of the study had leaked during the April conference in Geneva, Wanda published the details and undermined its conclusions with a scathing analysis, which infuriated her old friend Fuad Rouhani, who by then was OPEC's first secretary general. Wanda delighted in the notoriety.[15] Nevertheless, Wanda generally gave more favorable coverage to OPEC and its leaders than she did to the majors or the consumer governments; in 1962 alone, *PIW* cited Perez Alfonso in nearly half of the issues, often in lead articles—far more than anyone else. Wanda reported his objection to U.S. import controls, his desire to coordinate royalty rates among the producer countries, his hopes for Venezuela's new state-owned oil company, and his "indefatigable" efforts to build OPEC unity. When the United States tightened import quotas on Venezuelan—but not Canadian—oil, Wanda gave Perez Alfonso an entire page to express his outrage in his own bylined commentary. Venezuela, as America's oldest and most important supplier, he said, deserved better: "If this trend continues, we will be left with the bones—and we are not dogs."[16]

Even though Perez Alfonso had been angry with Wanda in 1960, *PIW*'s extensive coverage of OPEC pleased him. He met with her several times in 1962 and often cabled his latest policy statements in time for her Thursday deadline. When she visited him in Caracas in April 1963, he greeted her warmly and talked about his desire to live a simpler lifestyle. He then gave her one final scoop: he was about to resign, and he would cite poor health and exhaustion from constant travel as his reasons. The underlying cause, which she could not print, was his frustration at OPEC's lack of progress and his government's increasingly conciliatory relations with the United States.[17]

Wanda also kept in touch with OPEC's other founder, Tariki. They had repaired their friendship in 1961, so when she received a plaintive letter from him in June 1962, she decided to see him in Cairo, where he had retreated after he was replaced by Yamani in March. "He needs my advice and wants to talk to me," she wrote Ponikvar. She was not eager to go; she had other commitments in Europe, the airfare was expensive, and she was exhausted. Besides, she doubted that "there'd be any real copy (and Tariki won't reveal any OPEC secrets)," and it might get her into trouble with Yamani. But she decided to go anyway. When she got to Cairo, Tariki insisted on seeing her every day for a week. He was lonely, she was sympathetic, and she liked playing endless rounds of chess with him.[18]

Of course, she got a story from him, too. "The world of oil," she wrote, "has not yet heard the last of Abdullah Tariki, Saudi Arabia's outspoken former oil minister." He had resigned, not been fired, he insisted. He would continue to promote OPEC and seek "to improve conditions in Saudi Arabia" by writing a book on "our side of the story — what the oil companies haven't told." This explanation, however, left out some important information. Wanda protected her old friend by not delving into his conflict with Crown Prince Faisal and other members of the royal family, just as she had earlier that year when Tariki had "resigned." (In the late 1950s, he had become an important ally of the so-called Free Princes, a small group that called for a constitutional monarchy in Saudi Arabia.) Tariki had seen her twice in New York in early 1962 on his way to and from Caracas, just before Faisal forced him out. In March, she reported an inflammatory — and prophetic — statement of his: OPEC would eventually shut down production so that the majors would realize the "world cannot live without the Mideast's oil." But in that same issue, she had to insert a brief item at the last minute: Saudi Arabia's oil minister had just been "replaced." She added no explanation or comment. Her publication, she had decided, would report and analyze oil developments, not politics. Oil was controversial enough.[19]

OPEC alone gave her plenty of good material. After her week with Tariki, she returned to London and then New York, where she wrote favorably on the organization's latest efforts to confront the majors. OPEC's bid to renegotiate royalty payments because of a "small 'detail' of accounting procedure" was, Wanda noted, a "clever gambit" to gain more revenue. "All in all, it looks like an interesting chess game—and the negotiations should make interesting watching." And as many of her oil friends knew, Wanda loved to play chess.[20]

At the end of the first year, she tallied up *Petroleum Intelligence Weekly*'s successes for her readers: key decision-makers were quoting *PIW*, from the White House to Europe's energy planners, from Middle East ministers to banking officials. She underscored *PIW*'s speed: it reported Iraq's tax on tanker exports two weeks before other papers and published an OPEC memo "a full week before the dailies caught up." Some of its scoops—more than two dozen in 1962—were also cited in publications from London's *Financial Times* to Venezuela's *Carta Semanal*. Wanda had good reasons to celebrate, but, as was her wont, she remained nervous.[21]

ESPIONAGE AND JEALOUSY

In 1963 and 1964, two accusations that nearly derailed Wanda—one from OPEC, the other from the majors—showed how vulnerable she was to threats by institutions. Her survival, in turn, underscored the sheer strength of her reputation. OPEC's charge against her stemmed from Rouhani's frustration with the refusal of the majors and their governments to acknowledge OPEC's existence. Unwilling to meet or negotiate with any OPEC delegation, the oil executives would only attend bilateral meetings: the Iran consortium representatives with Iranians, Aramco representatives with Saudis, and so forth—just one of the ways the majors sought to forestall the organization's efforts to develop a unified challenge. On the surface, the majors pretended to be unconcerned, but their savviest executives knew that OPEC had the potential to become a serious problem.[22]

Indeed, some of them were so worried that they managed to get a secret waiver in 1962 to allow them to meet regularly. On their behalf, John ("Jack") McCloy, the lawyer-cum-diplomat extraordinaire, persuaded John F. Kennedy's administration (and its successors) to permit the five U.S. majors to meet regularly with Shell and BP by granting and renewing a waiver that guaranteed protection against the kind of antitrust law suits they had had to contend with in the early 1950s. McCloy argued that a foreign cartel that controlled 80 percent of the West's known oil reserves had serious implications for national security. From 1962 until the mid-1970s, the Seven Sisters' chief executives gathered alternately in New York and London for semiannual and then quarterly meetings, including detailed briefings from U.S. State Department and CIA officials. (Information on these meetings came out three decades later with the opening of McCloy's personal papers.) Despite their rivalries and different corporate temperaments, the majors thus developed a fairly coordinated approach to the OPEC challenge. "We did use the McCloy committee as a very valuable umbrella under cover of which we could talk," Sir David Barran recently recalled. "We really dug deeply into the issues." But when "we came close to the danger point of talking prices, Jack would warn us off." McCloy thus served as the U.S. government's enabler and referee for something the five American majors, under threat of antitrust prosecution, had been unable to do for decades. The informal oil club, thanks to OPEC, had finally found a formal way to meet and develop mutually beneficial strategies.[23]

And they managed to keep their club's meetings out of the press. Wanda was most likely aware of these gatherings, Critchfield speculated, but there is no specific evidence that she was. Her appointment diaries show that she often had lunch or dinner with British oil executives in New York at the time of McCloy's meetings, so it may well be that she knew but was bound by their insistence on confidentiality. She may have also appreciated what was for her a genuine national-security argument: that the majors were coordinating their response to the threat from new oil exports from the Soviet Union, a subject *PIW* had tackled. Nonetheless, publishing news about regularly scheduled confidential meetings of the Seven Sisters in Jack McCloy's offices would have been a major scoop.[24]

The brief influx of Russian crude on Western markets in the early 1960s did unsettle the majors because of its effect on the growing surplus of oil on world markets, but it also caused problems for OPEC. Its officials decided to set aside their demands for higher prices and to focus, instead, on getting higher royalty payments—a way to boost their revenue without a price increase. Since the companies would not meet with Rouhani as OPEC secretary general, he used his position as a senior Iranian official to start negotiations with the majors over the way they calculated royalty payments. After several inconclusive sessions in 1963 and early 1964, Rouhani became convinced that the majors were anticipating his every move because of leaks. "Someone is telling the other side!" he finally exclaimed. A likely someone was Wanda Jablonski.[25]

Rouhani did not publicly accuse Wanda of espionage, but he saw her as a prime suspect, one of his colleagues recalled. They did not know that the culprit was one of their own, someone revealing their tactics directly to the British. Undisputed evidence in recently declassified British documents proves that Reza Fallah, a fellow Iranian OPEC delegate and one of the most powerful officials in the NIOC, Iran's state oil company, was secretly relaying details of the organization's strategy sessions to British company sources. "I am shocked to see this information," said one of Fallah's colleagues when recently shown the documents. At the time, Fallah appeared above reproach. He was the NIOC's chief of refinery operations, a "brilliant, influential" engineer and administrator, the "best informed oil man who was close to the Shah." He "had to be doing it for personal gain," said his former associate. "The Shah never would have allowed him to do this."[26]

The evidence is substantial. "Dr. Fallah has been reporting secretly by telephone to the Consortium on developments," one British official wrote in a confidential memo in 1964. Fallah told OPEC associates that his frequent international telephone calls were "to obtain further instructions from his government in Tehran," but he was actually calling BP's Joseph Addison to brief him and get his advice. "OPEC has once again (thanks to the stand taken by Iran) shirked the decision to take unilateral action against the companies," another British energy official noted in a classified report. It was "the best outcome of the meeting that we could have hoped for."[27] Fallah managed to postpone an OPEC vote on a draft resolution by saying he needed further consultations with the shah, the Foreign Office noted. In reality, he quietly slipped into London "to clear his lines with the oil companies." As a result, BP and the Foreign Office were pleased that Fallah was then able to guide

OPEC to a "less hostile" outcome for the companies once Rouhani resigned from his post in mid-1964. To them, Fallah's success underscored the importance of Iran's decision to stay in OPEC as a way to moderate its decisions.[28]

Years later, Sir Eric Drake, Abadan's former refinery director and later BP chairman, recalled Fallah's role with some distaste, calling him "quite a crook" and "the most common sort of cur." Sir Denis Wright, then the British ambassador to Iran, also remembered Fallah as an "absolutely corrupt" BP man who became quite wealthy.[29]

Although Rouhani's suspicions remained private, Wanda's second major challenge in 1963 became threateningly public indeed: William Whiteford, the chairman of Gulf Oil, called Wanda to tell her that Helen Avati was having an affair with Ted White, her contact at Gulf's Paris office. Stunned, Wanda angrily replied, "Give me proof." When he gave none, she vigorously defended her reporter.[30] But then Whiteford struck a real blow: He canceled Gulf's *PIW* subscriptions and tried to get other majors to follow suit. Gus Long, Texaco's autocratic chairman, readily agreed. Frustrated by the depth and accuracy of *PIW*'s reporting, Whiteford and Long apparently hoped to use this incident to damage *PIW*'s reputation, or at least rein in its journalists. Avati was appalled: "I don't remember revealing Gulf secrets or that I had even written a story on information from Ted White. The charges were strictly that I was having an affair." There was no truth to the claim, but the effect was damaging. White no longer spoke with Avati and avoided her at oil gatherings. Like Wanda, Avati understood that rumors about affairs with sources were inevitable for female reporters. She worked in a man's world: all the oil executives and government officials she dealt with were men. But these particular charges and the response from Gulf and Texaco stunned her. There was nothing she could give to prove her innocence — except her word.[31]

Wanda believed her. "She stood by me," Avati later recalled. "She told me not to worry about it." Wanda had dealt with similar rumors herself, but this incident worried her. Would other companies follow suit? Completely dependent on subscriptions, she was vulnerable to threats to her reputation. Between Gulf and Texaco, she had lost more than forty subscriptions, more than 5 percent of her total. "What are we going to do?" she anxiously asked Buckley. Nonetheless, she resolved to press on. She would not be intimidated. During Buckley's seven years at *PIW*, he said, "Wanda never asked me to kill a story."[32]

Gulf and Texaco, as it turned out, had a harder time without *PIW* than vice versa. Gulf Oil executives sheepishly turned up at Kuwait Oil's headquarters in London to photocopy the latest issues, and Gulf's representative in Washington sought out the State Department's Andrew Ensor, saying, "I'm desperate. I need to read your latest *PIWs*. Just please don't tell my colleagues I'm doing this!" Texaco and Gulf executives in New York surreptitiously photocopied friends' issues and kept the latest tucked under other papers on their desks. "Wanda got so much mileage out of the cutoff," Buckley recalled. She told everyone what the two companies had done: "They tried to pull the rug out from under us but we're going to keep being as tough as ever," she would say. Indeed, publicity over the incident may have generated enough extra subscriptions to replace the ones lost.[33]

Texaco's chairman did not subscribe again for six years. "I remember when Wanda got a letter in 1969 from Gus Long asking to re-subscribe," said Marshall Thomas, by then *PIW*'s markets reporter. "He groveled with 'Dear Wanda, old pal' and so on—it was great." For someone who had held a grudge against Wanda since the 1950s, this must have been embarrassing, but Long enclosed a check for one subscription and signed it himself.[34] Wanda had stood her ground against two of the Seven Sisters—and won.

She relished smaller victories, too. At one point in the mid-1960s, Peter Walters, who later became chairman of BP, invited her to a company party at the New York Athletic Club, but she was denied entry because she was a woman. Walters enjoys telling the story: "'Look,' I told them, 'She's not a woman. She's Wanda Jablonski. If the queen of England were coming here, you'd let her in.'" The club conceded, "and then proceeded to treat her like the queen of England."[35]

CREDIBILITY FOR OPEC

OPEC was unable not only to get the majors to acknowledge its existence but also to get the mainstream Western press to consider it newsworthy. Before 1967, the *New York Times* had virtually nothing in its OPEC file, recalled its former energy reporter William Smith. At the *Wall Street Journal,* James Tanner and Roger Benedict gave consider- able coverage to the domestic oil industry, but OPEC was not yet of great concern to them. London's *Financial Times* and *Economist* paid more attention, but only intermittently—and they often cited *PIW.*[36]

Tariki's abrupt dismissal in 1962 and Perez Alfonso's retirement in 1963, however, marked a significant change in OPEC's prospects. Their passionate insistence on seeking greater equity from the majors faltered as the Saudi government, under Faisal's leadership, decided to pull back from all-out confrontation. As OPEC's principal founders, their impact was significant but relatively brief, as each lasted only about five years in power.[37]

Wanda's relationship with Yamani, on the other hand, lasted for three decades. In their first extensive interview, she publicized Yamani's call for "some kind of partnership" with foreign oil companies in new ventures, but Yamani worried about Wanda.[38] Because some OPEC of- ficials disliked her, "I was not totally at ease with her at first," he said. Still, he found her honest and frank. She wrote exactly what he said, "no more, no less." And she appreciated his effort to get the majors to nego- tiate with OPEC, not just individual countries. When he told her about bringing OPEC officials to what were supposed to be Saudi-only meet- ings with Aramco, she laughed and said, "What a great way to pull their legs!" That, Yamani said, was the start of a "real friendship." Every time he came through New York, he tried to see her, often visiting at her apartment. "You could learn a lot about [oil] people from talking to Wanda. She was a very good source of information—and reliable, al- ways reliable."[39]

Yamani appreciated Wanda's interest and hospitality, something he found lacking with some oil executives, such as Eric Drake, BP's chair- man in the 1960s. Drake himself recalled that when he once received a visit from Yamani and Abu Dhabi's young oil minister, Mana Said al- Otaiba, he found them "so green that I told them to go learn something about the oil business before they came back to talk with me."[40] Wanda, did, however, sometimes unsettle Yamani. In 1963, she upset him—and infuriated the oil companies—by giving prominent coverage to Tariki's call for the nationalization of all Arab oil operations at the fourth Arab oil congress in Beirut. She published Tariki's entire speech—all eight pages. Tariki used a wealth of U.S. statistics to show that Aramco had vastly higher profits than any other American company investing abroad. The next week, *PIW*—without comment—quoted Yamani as saying that Tariki, who had moved to Beirut to become a consultant, was not expressing the Saudi government's views. The fact that Tariki was out of power, with little or no prospects for a return, did not dissuade Wanda. His main point about Aramco's huge profits, and all the juicy details, was news, whether it angered the majors and Yamani or not.[41]

Her coverage of Iran's efforts to expand its industry expertise gave it added stature in the oil world. In the early 1960s, for example, if the NIOC wanted to buy steel pipe from a Western company, it had to "jump through hoops with lots of credit application forms" and then was still expected to pay in advance, Rostam Bayandor said. "Wanda helped put NIOC on the map." When the Iranian oil company made its first bid to attract foreign companies into joint ventures based on its own exploration work, top officials invited Wanda to visit. By giving her an exclusive story, they hoped Western oil companies would take them more seriously, and it worked. The NIOC collected huge cash bonuses for those deals. Wanda was "instrumental in getting the majors to realize that they had to get involved with us in new ways," Parviz Mina, the NIOC's international director, later recalled. She also wanted the producing countries to gain "the confidence we needed to take on these people." Mina's colleague, Farrokh Najmabadi, concurred: "It helped to have Wanda write stories that showed how skilled we were."[42]

Amir Abbas Hoveyda, Iran's prime minister in 1965, also appreciated her attention and made shrewd use of it. "Even I, as a retired oilman," Hoveyda wrote to her, "have found at times your articles a source of inspiration, a bone of contention, a cause for vexation and above all, material for quotation." That same year, he gave Wanda an exclusive interview in which he warned the consortium that Iran expected a 20 percent increase in output, prompting a flurry of telegrams among London, Washington, and Tehran. The coverage also helped attract independent oil companies, which signed contracts more favorable to Iran.

Some then made prolific offshore discoveries in the Persian Gulf, further challenging the majors' lock-grip on Iranian production.[43]

Wanda's apparent omniscience gave her legendary status in Iran. During a confidential NIOC meeting in the mid-1960s to determine which foreign companies would be awarded new exploration acreage, it was announced that she had arrived and was waiting outside. "The chairman of the committee, only partly in jest," Rostam Bayandor recalled, "immediately told someone to go and ask Mrs. Jablonski which companies had won."[44]

In 1965, Wanda attended the fifth Arab Petroleum Congress, in Cairo, where she reported on strident calls for nationalization: for the Arabs, the "state takeover of oil is only a matter of when and how." The majors' strategy for dealing with OPEC—seemingly endless, inconclusive negotiations—had exasperated the nationalists. "From now on," her Kuwaiti friend Feisal Mazidi said at the meeting, "we must legislate what we want, not negotiate." OPEC's chief economist, Francisco Parra, concurred, but warned: "We must approach this subject with ice-cold logic and not with red-hot emotion." For *PIW* readers in the oil industry, the warning was unmistakable. Indeed, OPEC's bid for power in the early 1970s should not have come as a surprise to them.[45]

Although the rhetoric in Cairo was public, Wanda uncovered a story in early 1966 that revealed how the oil nationalists were actually chipping away at the majors' control—a story that would have remained dormant, or at least misunderstood, without her. Kuwait's government was preparing for a showdown with the major oil companies, she warned. "An almost insoluble impasse" had developed between Kuwait and the Kuwait Oil Co., the BP–Gulf Oil joint venture. Contrary to the "widespread impression abroad," the controversy was not just a question of taxation and other payments. During several weeks of interviews, Wanda found that the real issue was state sovereignty, and said so bluntly. Kuwait's ruling family had not made a final decision, but prospects for a messy confrontation were high. Without naming names, she laid out the details of the conflict and explained why the repercussions could affect the entire Middle East.[46]

How Wanda got this story shows how complicated her task was as a reporter and consummate insider. She had come to Kuwait via London, as usual, where she first did her homework by debriefing executives such as Shell director John Loudon and BP's chief, Sir Maurice Bridgeman, along with other luminaries such as Frank McFadzean, managing director of Shell, and David Steel, later chairman of BP. Then, after a week of meetings and dinners with Kuwaiti sources and Kuwait Oil officials, she sought out U.S. Ambassador Howard Cottam for guidance on how to interpret the role of Mazidi, a key figure in the controversy and an advisor to Sheikh Jaber al-Sabah, the new prime minister and future ruler. Mazidi had revealed his strategy to her a year earlier, but she had kept it secret until he allowed her to disclose it in June 1965. By March 1966, his plan was coming closer to reality, but she wanted Cottam's view on whether Jaber shared "Mazidi's ambitions for 'cracking' the major oil companies."[47]

In a cable to Washington, Cottam aptly described Wanda's dilemma: "Miss Jablonski said Mazidi had confided in her so intimately that she could not tell me all of it and certainly could not publish very much of it; nevertheless, she had come to Kuwait to write the 'oil story of the year' and now was not sure she could do it. The plot was too thick and its architect too dangerous to the oil business for her to deal with." Mazidi had always talked frankly with her, she told Cottam; she sometimes helped him with "technical advice." His approach was different from Mossadegh's, she believed. He would proceed "legally." Cottam, however, was skeptical about how solidly Jaber stood behind Mazidi. His own assessment of Wanda is also significant: Wanda appeared to be part of a team of advisors for Mazidi—or "at least Mazidi seems to take her advice."

In July, Cottam again cabled Washington to warn that another *PIW* article would be the "likely opening of campaign to pressure companies into acceding to major [Kuwaiti government] demands." In the article, Wanda outlined the arguments from all sides and then spelled out how far each side could stretch without a breakdown. It was vintage Wanda Jablonski. As Parviz Mina explained, "She was instrumental in giving warnings. She'd say if you go beyond this point, they might break and provoke a real crisis. That was very helpful to us."[48]

With articles like these, "Wanda helped speed up the process" of

OPEC countries seeking more control, said Aramco's Mike Ameen. "She identified the pressure points." Shell's David Barran concurred. Some oil chiefs criticized her for these exposés, but "I never heard anyone accuse her of misreporting the truth." Yamani said that Wanda's role was critical because she "gave legitimacy in her articles to OPEC's views when the oil companies wanted to pretend OPEC didn't even exist. She gave a voice to OPEC."[49]

But this time, the Kuwaiti sheikh, by delaying a decision, was overtaken by events. In June 1967, the Arab-Israeli war forced the oil nationalists to retreat, at least temporarily. The Arabs not only collapsed on the battlefield within days of Israel's preemptive attack, but their first attempt to brandish the "oil weapon" failed badly. Just after fighting began, five Arab members of OPEC declared an oil embargo against states friendly to Israel—primarily the United States, Britain, and Germany—but this partial embargo boomeranged. With the help of official and unofficial networking, the oil companies drew on spare U.S. production capacity, reshuffled tanker deliveries, and boosted exports from non-participating Iran and Venezuela. In effect, Iran took revenge on the Arabs for their lack of support during its aborted nationalization attempt in the 1950s. It also continued to secretly supply Israel with oil, as it had been doing from the time of Mossadegh in 1951 and would continue until the 1979 Iranian revolution. As a result, the banned countries did not suffer any real oil shortages.[50]

Hampered by this lack of unity, OPEC did not appear to be any match for the majors, yet in many ways it exerted more power in the 1960s than it has been given credit for. That OPEC prevented the majors from further cutting posted prices, given the persistent glut, was significant. As *PIW* pointed out, OPEC caused oil-company profits on Middle East crude to decline in percentage terms as market prices fell, while producer governments' incomes—set by fixed posted prices— gained. But as Barran noted, "Wanda was one of the first who realized that OPEC had, to a considerable extent, got off on the wrong foot." She helped the producer governments understand that "what was most important was price," not the distinction of having the highest level of actual oil production. This point was particularly true for Iran.[51]

LIBYA: THE DAM BREAKS

The rapid failure of the 1967 Arab oil embargo reinforced the oil club's power but also underscored OPEC's disunity. By supporting international oil companies rather than its fellow OPEC members, Iran and Venezuela exacerbated long-standing rivalries. Earlier initiatives aimed at challenging the majors, such as the one Wanda reported from Kuwait in 1966, withered. The industry's boost from U.S. domestic production during the 1967 crisis—nearly a million barrels a day—was virtually gone by the end of the decade, as those still-regulated old fields ran out of surplus capacity and demand continued to soar. It would take a newer member of OPEC, Libya, to figure out the oil club's Achilles' heel and galvanize other OPEC countries into action. And when it did, Wanda had an experienced reporter in Tripoli to explain the significance.

For years Wanda had followed Libya's development as an oil producer. She first went to Libya in 1960, landing in Tripoli the day after Christmas during a rain. She was exhausted from nearly four months of nonstop travel through the Middle East, yet she pushed herself to go to Tripoli because of the buzz she was hearing from oilmen. Libya had not yet exported a single drop of crude, but it was a "big play," and for oil-hungry Europeans, its fields lay only a short trip across the Mediterranean. Wanda returned again in 1963 and again in 1965, each time lining up exclusive interviews with whoever was the oil minister. Since North African oil exploration was up for grabs, presidents of the independent oil companies (such as Conoco, Marathon, and Phillips), eager for Wanda's latest intelligence from the region, often stopped in New York to pick her brain on their way to North Africa. In return, Wanda got exclusive information about all the competing companies.[52]

Libya's production potential was sky-high, and so was its potential for trouble. Tariki found work there occasionally as a consultant, and he certainly could stir things up. Another friend from Wanda's visit to Abu Dhabi, Tim Hillyard, was leading BP's exploration team in Libya, and she quizzed him repeatedly on Libya's prospects. Tariki and Hillyard both knew her detective talents, yet the story she published in 1965 surprised them both. Piecing together information from company and government sources, she showed that the majors' net profits on Libya's then rapidly growing exports of crude amounted to nearly 30 percent on each barrel sold, compared with less than 10 percent for the independents. The story was a gold mine, replete with exact figures on the majors' tax payments and production costs. Here was detailed proof of the majors' extraordinary profits. Printed in the authoritative *Petroleum Intelligence Weekly*, this information undermined the majors' stance, opening a chink in their armor to attacks from critics in Libya and from independent oil companies trying to break the majors' stranglehold on Middle East production. This scoop was Wanda at her best.[53]

She had a feeling that Libya would become important. "Part of Wanda's genius was her ability to foresee potential developments," said former *PIW* and wire service reporter Ken Miller. He arrived in Libya in June 1965, the day before a major story broke on exploration concession agreements. "It was unbelievable. No one else was there to cover it." Wanda's prescience "became part of industry folklore," Miller recalled. "That exact timing was just luck, but it was brilliant that she sent me in." From 1965 until 1971, Miller filed one exclusive story after another from Tripoli, with sources ranging from the high-flying independent oilman Bunker Hunt, who had his hotel suite decorated "like a Texas bordello," to Libya's oil minister, Fuad al-Kabbazi, a cultivated, Italian-educated poet. "Kabbazi welcomed me as a sort of intellectual companion," Miller said. "It was hard for him to talk poetry with the young kids in his ministry, like the undersecretary, who had lived his childhood without shoes."[54]

Miller's friendship with Kabbazi proved fortunate. Shortly after the Arab-Israeli war of 1967, strikes and riots broke out in Libya, with Americans particularly under attack. Oil company personnel sought refuge at a U.S. Air Force base, but Kabbazi, who by then had been replaced as oil minister, invited Miller and his wife, Leyna, to live in his home. Miller reported from there for several weeks, even managing to get breaking news to *PIW* through the government's censors by writing his cables in code. To signal that the entire Libyan cabinet had quit, he reported, "Whole kit and caboodle uphang suits." Because of Miller's exceptional vantage point, the White House and a number of oil companies called *PIW*'s New York office regularly for news from his cables until the crisis calmed down a few weeks later.[55]

But of all the exploration deals signed in the 1960s, it was Occidental Petroleum's 1965 concession contract, the one that Miller covered in his first story out of Libya, that proved the most significant. Dr. Armand Hammer's upstart independent oil company, which won drilling rights to acreage that had been unsuccessfully explored, struck "pay dirt" in 1966 with a discovery so prolific that Libya's production more than tripled in four years, going from impressive to meteoric.[56] As that country's oil output and wealth grew exponentially, so did its resentment at what it saw as the Western oil companies' colonial control. Libyan officials kept trying to chip away at the companies' power, but largely in vain. As Miller chronicled their efforts, his scoops were often cited in U.S. embassy cables. Then, on September 1, 1969, a group of young army officers, led by the dynamic Muammar al-Qaddafi, deposed the ailing King Idris. A short time later, Qaddafi invited the OPEC world's most controversial consultant, Abdullah Tariki, to advise his staff.[57]

Qaddafi's strategy soon became clear. Unencumbered by OPEC's sense of failure with the 1967 oil embargo, he first tried to get a price increase from the majors. When they resisted, he put pressure on the most vulnerable oil producer—Occidental, which had no substantial production outside Libya. Armand Hammer appealed to Exxon, the major with the largest production in Libya, to help him form a united front, as the majors had during Iran's nationalization bid in the early 1950s, but Exxon's chairman turned him down. As one of Hammer's advisors noted, Exxon did not want to rescue a "non-fraternity type" who had so immoderately driven up Libya's production. And so, under threat of a total shutdown, Occidental caved in to Libya's demands, agreeing to a steep increase in taxes and in the posted price. With that, the long-standing fifty-fifty profit-sharing ratio that had dominated the oil world for two decades collapsed. Occidental agreed, in effect, to give Libya 55 percent in September 1970. Suddenly aware of their mistake, some majors tried to band together in a common front, but it was too late. Libya would negotiate only with individual companies. Within six weeks, other majors—Exxon, Mobil, Texaco, Chevron, and BP—also capitulated to Libya's terms, one by one.[58]

For the first time, an OPEC country had single-handedly imposed its will on the major Western oil companies. Indeed, the more radical countries within OPEC—Libya, Algeria, and Iraq—had the companies so worried that they suddenly looked to OPEC, the organization, as a potential moderating force. "A profound change" had occurred, *PIW* reported: "The oil companies are now convinced that OPEC alone, as a body, holds the key to 'stability' in world markets." After years of pretending publicly that OPEC did not exist, and reacting smugly to the Arabs' futile effort to use the oil weapon in 1967, the Seven Sisters found the tables had turned. Just as startled at this change in fortune, OPEC's leading members—Saudi Arabia and Iran—could hardly believe Libya's success. Only three years earlier, the Arabs had been humiliated by the collapse of their partial embargo; now, in a matter of months, Libya had brought all the leading oil companies to submission. The oil club's "mystical power," as the shah had called it, was beginning to evaporate.[59]

BACKING OFF ON THE HOME FRONT

For the first five years of publishing *PIW*, Wanda kept up a hectic pace. Her frequent travels brought *PIW* many of its best stories. And the newsletter was, indeed, influential, said Mobil's chairman, Rawleigh Warner, who read it every Monday. "*PIW* would make everybody mad but everybody had to read it," recalled BP's Walters. "There were stories particularly about industry secrets or trends that we'd been trying to keep quiet and *PIW* would blow them wide open." Sometimes it even moved the stock market, as the *New York Times* noted in 1969. PIW often beat the competition, the *Wall Street Journal's* oil reporter, James Tanner, later said, because of contacts just below the board level: "I couldn't talk with that middle echelon the way *PIW* did—that's how they got so many scoops."[60]

Buckley and Avati excelled at this kind of reporting, but in late 1963, Wanda decided abruptly that Jack Hartshorn was too expensive, too long-winded for her readers, and not producing enough. "That was typical of Wanda to suddenly go cold on someone," Buckley recalled. "There was no real reason for her to fire him," although, in addition to writing for the *Economist*, he was also completing his first book. Years later, Hartshorn insisted he did not have a fight with Wanda. "Our styles were just different," he said. Remarkably, they remained friends; but to his wife's dismay, Hartshorn returned his 5 percent stake in *PIW* without asking for—or receiving—any compensation.[61]

Wanda was not a good manager of those who worked for her. Particularly as she got older, she grew more impatient and imperious. Buckley knew his days were numbered when Wanda finally persuaded her *Petroleum Week* colleague, Georgia Macris, by then bored with *Oilgram News*, to join *PIW* in 1967. Macris was furious when she learned that her hiring led to Buckley's departure; she had looked forward to working with him. But within a year of becoming managing editor, Macris herself issued Wanda an ultimatum. Two cooks in the kitchen would not work; one had to get out. So Wanda got out. "I've always been told I don't know how to handle people," Wanda told Macris, "so I don't want to deal with them."[62]

From 1967 onward, Wanda began a transition away from full-time reporting. She finally had the financial stability she had long sought: subscriptions exceeded nine hundred, *PIW*'s reputation was solid, and she had assembled a substantial staff. Tired of traveling, and worn out, she no longer had enough "fire in the belly," Macris said, but she also did not have the patience to deal with the daily demands of editing and staff management. What's more, she had gotten "too close to the people she was hurting," recalled a former Shell executive and *PIW* reporter, John Bishop. As publisher, she could distance herself somewhat from *PIW*'s controversial stories, claiming she "could not control what her staff did." So she backed off.[63]

Many on the staff flourished once Wanda left the paper's day-to-day management to Macris. Avati continued to report exclusives on Europe and the newly emerging Algerian oil fields. Ken Miller signed on in 1965 to cover OPEC and the Middle East from Libya. Margaret Clarke, a demanding but fun-loving manager, agreed to move to New York to become chief copy editor. And Marshall Thomas, who would become *PIW*'s markets reporter for nearly two decades and then briefly executive editor, started in 1969. Although Wanda remained *PIW*'s sole owner and publisher for three more decades and kept a tight rein on financial and personnel matters, she only occasionally delved back into reporting.[64]

By 1969, Wanda was in her late forties and eager for a change. She decided to focus on her personal life. She redecorated her Central Park South apartment and bought a Long Island waterfront house near Oyster Bay for weekend entertainment. But rather than finding a satisfying personal life, Wanda suffered further losses and frustrations. In 1972, her mother became ill and died from pancreatic cancer. Although Wanda had chafed for years from her mother's critical comments, she suffered as her mother lay dying. "She said she came to the realization too late," one friend said, "that her mother was the most important person in her life, the one person who had really dedicated herself to her, who really loved her." But perhaps this comment came more from a sense of Wanda's guilt for her years of detachment than from a deathbed change of heart, since other friends said Wanda was not close with her mother. Eugene Jablonski lived until 1975, and Wanda occasionally invited him to her Oyster Bay home with some of her colleagues. Although Margaret Clarke remembered fondness between daughter and father, Georgia Macris did not. "I never saw her show real affection with anyone in all the years I knew her," she said.[65]

The affection she did seek from others would disappoint. Wanda's lack of close, long-term friendships was not for lack of trying; she simply could not maintain them. Increasingly self-absorbed and intolerant of human frailty, Wanda had been so absorbed in her career for so long that she did not have the skills to build and keep personal friendships, even though she maintained many successful professional ones. There were rumors about liaisons with executives at the majors—"Wanda in those days was pretty sexy," Mike Ameen recalled—but she had no lasting romantic relationships. She dressed fashionably—often in black and white, with metallic jewelry—and yet she was self-conscious enough to have plastic surgery on her nose and eyes in the 1970s. She liked to be in control, she told one friend, and she did not like relationships that gave men power over her.[66]

The most important liaison she had was with an old family friend, Brandon Grove, and it lasted for a couple of years. Grove's wife, seriously afflicted by Alzheimer's, was confined to an English nursing home, and he relished his time with Wanda when she visited London and when he went to New York. Between 1966 and 1969, he spent a number of weekends alone with her at her Oyster Bay home. "He certainly had a fondness for her—even a crush," his son recalled. In meticulous but brief daily diaries, Grove kept note of a familial relationship with Wanda at Oyster Bay; he spoke of mowing her yard of one acre, cooking live lobsters or *moules marinières*, cleaning the dishes, and mopping the floor. Wanda and Grove, fortified with "abundant libations," spent many afternoons and evenings playing chess, which Wanda won consistently. But Grove's son said he did not believe the relationship became "a full-fledged affair."[67]

There were casual relationships with other men during that time, including a handyman named Jim and a tall, elegant German. Erich Schliemann, whose German oil company was bought out by Texaco, spent a weekend with Wanda at Oyster Bay in 1967, and she once vacationed in the Bahamas with him. "We liked each other and laughed a lot," Schliemann recalled years later, although sometimes he would get "a little annoyed" at the way she "favored the Arabs." As for their relationship, "we didn't take things too seriously," he said, without denying they had an affair. Wanda, he claimed, "had a soft spot for him."[68]

Frank McFadzean, the acerbic chairman of Shell Transport and Trading, also courted Wanda for some time in the 1970s. The two met while waiting for a London flight. "So you're *the* Wanda Jablonski?" he asked. "You don't look like the bitch I thought you'd be." (At least that was her recollection of the encounter.) He gave her a huge diamond ring, and she may have even been engaged to him, according to several friends, including Yamani and his wife. Wanda liked McFadzean's irascible nature but could not marry him, she told a friend flippantly, because she would get bored living on the Thames as Lady McFadzean.

Some friends thought Wanda had married more than once; indeed, she told some she had been married twice, others three times. But just as she allowed an air of mystery about possible affairs, she enjoyed telling embellished stories about her love life, including an elopement with a young Slovak whom her family forced her to disavow, and a champagne-induced one-day wedding to an oil executive, which Margaret Clarke disputed, saying: "If you knew her well, you knew she didn't like champagne. And if you knew her well, you also knew she was a big tease." Jaqua, her executor determined, was her only husband.[69]

In her later years, Wanda's personal and professional friends said she was often lonely, even deeply so. She had spent many holidays away from her parents when they were alive, but now she felt especially bereft of family during holidays. She was a witty guest, said New York oil analyst Jim Hunt, who invited her to join his family for Thanksgiving for many years, but she did not connect emotionally with people. "It's not surprising that she never married again," Hunt said. "No man would ever be able to put up with her twenty-four hours a day, seven days a week." To Hunt, she was the scientist of the oil industry: "She may not have been able to dissect a frog in college but she dissected people like they were lab animals." Of course, that skill worked better on the job than in a friendship. In her later years, she had a succession of friends, both men and women, who did not stay close to her for long.[70]

Wanda's deep-seated anxiety and insecurity—the personal demons she had struggled with since childhood—became more evident as she became better known. With advancing age, the tracery of nerves beneath her skin became more visible; her legendary wit, increasingly mordant. Because she did not work regularly, she tried to find people who would play chess, backgammon, or bridge to feed her competitive spirit, but her friends who were not oil contacts—Eleanor Schwartz, Harriet Costikyan, and Beverly Parra, for instance—did not remain friends. She was too ill-tempered. Each had her own story of how she finally decided to stop seeing Wanda. Wanda could be so demanding, and even cruel, that the cooks and housekeepers she hired never lasted long. Only her secretary, Agatha Sangiorgio, who, at Wanda's insistence, had to work under the simpler name of "Agnes Stewart," had the fortitude to survive in long-term service to Wanda. She held her job for more than two decades.[71]

Wanda did continue to cultivate her oil network. Her old friends Perez Alfonso and Tariki were no longer in power, and some of her close friends among the captains of the oil industry had retired. But she maintained relationships with the next generation of leaders, with certain key OPEC leaders such as Yamani and Kuwait's new oil minister, Ali Khalifa al-Sabah, and the oil company chiefs, talking with them at oil industry conferences or visiting them at their vacation homes.[72]

More important for Wanda's readers, the most substantive explanations of the changes affecting the oil industry continued to be found in

Wanda with Saudi Arabia's Sheikh Ahmed Zaki Yamani and Mobil president William Tavoulareas at Yamani's home outside Taif, Saudi Arabia, most likely in August 1981.

the pages of *Petroleum Intelligence Weekly*. For years, Wanda and her reporters had exposed and explained the inner workings of the oil companies and their trading practices: production rates, profit margins, royalty payments, income taxes, freight costs. Then world oil supplies tightened as U.S. production hit a ceiling, Russian exports began to dry up, and new prospects in Alaska and the North Sea encountered delays—all of which *PIW* reported in detail. From the late 1960s to the early 1970s, the market shifted from a buyer's to a seller's market, and although this reflected more a transportation crisis than an actual shortage of oil, it was the short-term perception that mattered. *PIW*'s scoops and analyses were not only influential for oil companies and governments in both consuming and producing countries, and even in the Soviet Union, but they were also used by young Libyans—with help from Tariki—as tools to drive a wedge into the oil fraternity and make it crack. As one Kuwaiti told oil historian Anthony Sampson, "We thought we were pygmies facing giants. Suddenly we found that the giants were ordinary human beings; and that the Rock of Gibraltar was really papier-mâché."[73]

9.

Final
Confrontations

Georgia Macris looked up when she heard the teletype machine start to clatter, the signal that a story was likely coming in from overseas. It was press day, Thursday, October 5, 1972, but it was after 5 PM—too late, really, to fit in another article for the Monday issue, she thought. Cigarette in hand, Macris hunched again over the manual typewriter to finish editing a story in her cramped office. Despite *PIW*'s considerable profits after a decade in print, Wanda had not improved the dingy working conditions.

Macris then heard excited voices around the machine. Roger Benedict, the reporter whom Wanda had just hired away from the *Wall Street Journal*, was trying to read the cable out loud as the paper spewed forth, when suddenly he hollered, "Wow, this is the greatest scoop!" The last-minute file from Helen Avati was indeed dynamite—a story that would stun the oil world. Macris had to put it on page 1. And because she realized how sensitive the information was, Macris issued an order: the staff was not to utter a word about the author or its provenance. This was the biggest story of Avati's career, but she would not—she could not—take credit for it. If she did, everyone would know the story came from Paris, and she would have jeopardized her source. That she would never do.[1]

The story had all the important details of a secret agreement, signed

that very day in New York City: the long-awaited "participation agreement" between five Arab oil-producing countries and many of the leading international companies. The deal had been the central focus of
the oil world for months, and the secrecy about the negotiations had
been extreme, even by the industry's notoriously stringent standards. Although Libya had won the first showdown with the oil companies in
1970, and the OPEC countries had forced on them a cascade of com-
promises in 1971, *PIW*'s scoop on the specifics of this 1972 accord proved
decisively—five years after the failed oil embargo of 1967—that OPEC
had trumped the oil club. Not only did the details in the article prove
that the Arabs had won, but the timing of it could make matters much
worse for the companies. That the oil club was unable to keep the deal
secret—not even for a day—could undermine the entire agreement.
The Arab states might now do something drastic.

Thursday evening, while Macris was editing Avati's story, she called
a high-level Mobil source at home to clarify a few points and test the
way the majors would react to her news. She got stunned silence. Soon,
however, the phones were ringing with Aramco executives "all fuming
about the story," Macris recalled. "It was a huge embarrassment. They
were quaking in their shoes, thinking all hell would break loose." Yamani had given "strict instructions that no word of [the deal] was to
come out, and they assumed he would blame them for the leak. It was
the speed with which we got the story, not so much the details, that mattered most. There should have been no way for us to get that story." Was
there a mole in their midst, or had Yamani leaked the story just to set the
companies against each other? No matter what these executives asked,
Macris would not help them, but they, in turn, confirmed so much
through the questions they asked. For Wanda, the scoop proved to be
spectacular, but *PIW*'s future was once again in peril, because much
was at stake.[2]

"SO MUCH FOR SECRECY"

In early 1972, the oil world had nervously watched and waited as the
Arab producing countries and the majors negotiated the core issue—
the actual ownership of the oil, which was the companies' ultimate

source of power. "Participation" was Sheikh Yamani's euphemism for moving Saudi Arabia toward the complete nationalization of its petroleum industry but in a gradual way, since his country still needed Aramco for its technical expertise and market outlets. Yamani persuaded OPEC's four other Arab countries to let him take the lead in negotiating a general revision of the concession contracts. For many months, the companies—the Aramco partners, plus others such as British Petroleum and France's Companie Française des Pétroles (CFP)—dragged their feet, but as market conditions deteriorated and the more radical OPEC members (Libya and Algeria) escalated demands on their separate contracts, the companies finally decided that participation was better than immediate, outright nationalization.[3]

In March 1972, when the companies tentatively agreed to cede ownership of 20 percent of the concessions, *PIW* underscored the significance of this first step. "There's no doubt left now that a profound change is shaping up for the future structure and operations of the international oil industry," it said. The newsletter reported on declining U.S. production, a further tightening of global oil supplies, and rising prices on the open market. And it had some additional ammunition for OPEC's argument that prices should be higher: *PIW*'s technical analysis showed that oil import costs for Europe and Japan, adjusted for inflation, were less in 1972 than they had been in 1957. The negotiations then stalled over disagreements on payment terms. In July, a frustrated Yamani issued an ultimatum, sternly warning the companies to quit their "delay and procrastination."[4]

Without a glut, without alternative sources of supply, without Western government intervention, the majors had few options. Led by Exxon, the companies finally initialed the participation agreement on Thursday, October 5—*PIW*'s press day—in New York. At a press briefing the same day, Yamani said that the two sides had reached a tentative deal but that he could not reveal any details because he was leaving immediately to seek final approval from King Faisal and the governments of Kuwait, Qatar, Abu Dhabi, and Iraq, which might take several weeks. Yamani had sworn the companies to secrecy. They could not even tell the U.S. government (not that they wanted to). But the accord, as a confidential State Department cable noted, was expected to be along the lines previously announced—an initial 20 percent for the Arab countries.[5]

Because of this intense secrecy, *PIW* shocked the oil world with its Monday issue. The Arab states' initial "participation," the unsigned article said, would "start with 25 percent in 1973, not with the long-discussed 20 percent minimum." Symbolically, that extra 5 percent meant a lot. The OPEC countries had gained so much clout so quickly that they could keep setting and resetting their terms—and win.[6] Moreover, the leaders in these negotiations were not radicals. Saudi Arabia was the most conservative and most powerful player within OPEC. To the oil world, the bottom line was evident: the elite club of white Western men that had so long ruled the business was no longer in control. This recognition came a full year before the 1973 oil embargo, a full year before the rest of the world realized that the oil club had lost. That loss had far-reaching consequences. For the OPEC countries, it led to extraordinary economic and political power; for oil consumers, it meant skyrocketing energy prices, a severe recession, and unprecedented worries about energy security.[7]

The oil world assumed that the author of the story was Wanda Jablonski. To this day, Helen Avati will not divulge from whom she got the story, or exactly how. She is still bitter that Wanda disclosed seven years later that the story had come from Paris. "Everybody was furious with us," Wanda told the *New York Times* in 1979. "Zaki [Yamani] thought we got it from the companies, and the companies thought we got it from Zaki." She confided, "I'll tell you something, we got it from the French. We just asked ourselves, 'If a deal has been made, who else would have to be informed?'" Indeed, Avati had exceptional sources at the French company, CFP, and someone there most likely gave her a copy, Macris surmised, since Avati "had the exact wording, everything." Aramco had to keep CFP informed because the agreement affected Iraq, whose concession included CFP. France's state-owned company, however, had always resented the U.S. and British majors because they did not consider it their eighth "sister."[8]

Shortly thereafter, leading oil executives met again in London to reassess their negotiating options as part of the London Policy Group—a government-sanctioned gathering of the majors and independents involved in the Iran consortium. At one meeting, as senior executives gathered in a BP auditorium, someone tapped on the microphone and said, "Testing, testing. Wanda, are you listening?" But *PIW*'s 1972 scoop was no laughing matter for either side.[9]

The *Economist* relished *PIW*'s success. The agreement, it noted the following week, "has so far been most notable for the way in which a surrounding veil of secrecy was torn to shreds by the publication, within days, of extensive details in *Petroleum Intelligence Weekly*. Officially, these details are still denied, but they are true." The *Economist* summarized the article, citing *PIW*'s estimate of the industry's huge loss of profits, and concluded, "So much for secrecy."[10]

As for Wanda, she had arrived in Beirut on October 7 with Malachi Martin, a former Jesuit priest with whom she had just taken a romantic trip on the Nile. Approached several days later by several reporters who wanted to know how she had gotten such an amazing scoop, Wanda looked wide-eyed and answered, "I don't know what you're talking about." The reporters thought her story about floating down the Nile was typical Jablonski deception, and her relationship with Martin only added grist to the Beirut rumor mill. Ian Seymour, a reporter friend from the *Middle East Economic Survey*, later said, she "wanted everyone in her world to know" she was with Martin. They were "obviously lovers and quite taken with each other," Seymour recalled. "But I don't think I have ever seen a couple quite so mismatched."[11]

With the news of *PIW*'s hot story, Wanda left Martin in Beirut and flew to Kuwait, where Yamani and other Arab state ministers had gathered to discuss the agreement. While the ministers were meeting, she went with Seymour to shop at the *souk* for the afternoon. "When we got back to the hotel in the early evening," he recalled, "there was a cable for her at the desk. She went deathly pale—I can still see it in my mind."

"Ian, I need a drink," she said. The telegram from Georgia Macris said that Exxon had canceled all but six of its subscriptions and was trying to get its Aramco partners to do the same. "She was shattered," Seymour said. "She didn't know whether *PIW* could survive." Since Wanda could not face the restaurant, Seymour ordered room service and a bottle of whiskey. Perhaps unaware of her sometimes contentious relationship with Exxon's Howard Page, Yamani thought Page (although semi-retired) had slipped her the agreement, Seymour recalled. "Yamani had come on strong with Exxon about not leaking and he wielded a big stick," he said. "Exxon must have been in terror."[12]

Wanda had to respond. Her newsroom was "urgently awaiting any tactical suggestions you may have." Refund them the balance on the canceled subscriptions, Wanda telexed in reply, but continue to send them the issues until the subscriptions expire: "Tell them we honor our contracts." It was a phrase George Piercy, who had recently replaced Page, had, as Exxon's lead Middle East negotiator, often used in putting down the oil nationalists for trying to revise their contracts. Macris signed a check for $36,000 and held her breath, waiting to see who else would cancel. As BP's Peter Walters said, "All the majors thought about canceling our *PIW* subscriptions then. There was tremendous finger-pointing between the companies."[13]

Shortly after Wanda sent her reply, Yamani approached her in the Hilton lobby and found her holding a piece of paper. "What's that?" he asked. She showed him the telegram from Macris. He read it and replied, "But I liked your answer." What did he mean? she asked. How had he seen it? (Yamani apparently had someone, perhaps a hotel staff member, show him a copy of her telex.) He simply grinned at Wanda and said, "I have my ways."[14]

He offered to take her with him on his plane the next day to Riyadh, where he would meet again with Aramco executives to negotiate final points in the agreement. She agreed. Upon their arrival on October 17, Exxon's Piercy was at the airport to greet Yamani. When Wanda stepped off the plane with the minister, Piercy looked aghast. "I was carrying my briefcase and a bag, and Wanda had her luggage," Yamani recalled with a laugh. "She was carrying a feminine bag." So he said to Piercy, "George, it's too bad my hands are full, otherwise I'd carry the bag for the lady." Nonplussed, Piercy took her bag and accompanied them to the waiting limousine. Piercy, Yamani said, did not have a sense of humor; nor did he like the fact that Wanda would be visiting Yamani at his mountain home in Taif, near Jeddah, with her friend Malachi Martin.[15]

Martin was an erudite, gregarious, quite worldly ex-priest who had already published several books when he met Wanda through friends on Long Island. She was smitten, and the two traveled together to Europe and the Middle East soon thereafter. "Don't be surprised if I come back married," Wanda told Macris before leaving in September. After a week in Dublin and then Rome, Wanda took Martin to Egypt, where they spent nine nights at the Nile Hilton and saw some of her oil friends before taking a cruise to Aswan. It was like a homecoming in Cairo, she later wrote her friends Wadad and Mahmoud Makhlouf in thanking them for their hospitality. She asked them to give her love to a mutual friend, "Pussy," and "tell her I'll forgive her for trying to marry me off to Malachi—though I doubt he'll forgive her for that!"[16]

By the time Martin rejoined her at the Kandara Palace Hotel in Jeddah, it was the month of Ramadan, a time Wanda particularly enjoyed. Ramadan fit her rhythm: stay up and talk most of the night, then sleep and, if need be, fast during the day. Yamani invited Wanda and Martin to his home for many such evenings, but he grew increasingly puzzled by her interest in Martin. Yamani found him to be overly religious and rigid, whereas Wanda was not religious and liked to laugh. She was witty and fun. At one point when the three were talking, Wanda said, "Zaki, when are you going to take me to Mecca?"

"It's very easy," was his reply. "All you have to do is convert."

That was fine with her, she said. She would like to convert to Islam. "This got the priest so upset with me," Yamani recalled. "'You shouldn't talk like this!' he told me." Later, Wanda thanked Yamani for not taking offense and apologized for her friend. Whether that contretemps proved the end of their relationship is not clear, but immediately thereafter, Wanda took off without him for five weeks of whirlwind reporting. Upon her return to New York in late 1972, "a very angry" Wanda told Macris "she had washed her hands of Malachi." Something went wrong on the trip, Macris surmised. "It was a brief but fiery romance."[17]

What Macris did not know—nor did Wanda—was that Martin had a reputation for seducing women. An Irish-born charmer with a doctorate in archaeology, this Jesuit had taught at the Vatican after publishing a book on the Dead Sea scrolls and become the aide to an influential prelate. During the heady days of the Second Vatican Council in 1963, Martin dazzled the foreign press, which turned to him for insights and quotes. He also had an affair with the pregnant wife of *Time* magazine correspondent Robert Kaiser, who then discovered the priest's other liaisons. In 1965, Martin turned up in New York, where he eventually became a media commentator on the Catholic Church. When Wanda met him in the early 1970s, he was working on a book about Vatican II. But he was also growing obsessed with Satanism, and he eventually became an advocate of exorcism.[18]

Whatever the cause of their breakup, Wanda left Martin and traveled alone to Riyadh in November. She got exclusive interviews with Abdulhady Taher, chief of the new state oil agency, Petromin, and Prince Saud al-Feisal, the deputy oil minister, who said his government wanted to invest in refining and marketing outlets in oil-consuming countries. He was eager to learn about the business: "When you realize you'll be responsible for selling something like 10-million barrels daily within ten years, you've got to get on your horse fast." Although Wanda packed her bylined commentary with news, she also packed it with compliments. The prince, she wrote, combined an "innate grace and dignity, with statesmanship and candor, modesty and intelligence." Wanda quickly realized she had gone too far. "I'm sorry I put in quite so much praise—in cold print," she wrote a friend in Jeddah.[19]

She then stormed around the Middle East that November and early December, filing stories and in-depth interviews from nearly everywhere she went. From Riyadh she flew to Dhahran, then various emirates on the Persian Gulf, Oman (for the first time), Baghdad, Kuwait, and finally back to Beirut to debrief with friends.[20] She was particularly drawn to Abu Dhabi's personable young oil minister, Mana Said al-Otaiba, who told her that, rather than relying on foreign experts, he wanted his own people to sell the emirate's share of production and invest the proceeds in refining and marketing outlets abroad. He became one of "my boys," as she called them—rising young stars from oil-producing countries who sought her advice and instruction and visited her in New York or Long Island.[21]

What had suddenly spurred Wanda into action? She had not done this much reporting since 1967. Surely the upheaval in the oil world had fired her up, perhaps anger at Malachi Martin played a role, and probably she envied Helen Avati's success. Perhaps Wanda felt the need to show her staff—and perhaps herself—that she was still a master of the game. Wanda was jealous of Avati, a "sensational reporter," Macris later said. But given the Exxon cancellations and the possibility of more, the determining factor must have been her need to ensure that *PIW* would survive. Although Exxon had not deposited *PIW*'s refund check, word went out that its publisher was banned from the Exxon building. More than anything, Wanda was determined to prove to Exxon and the oil world that they could not do without Wanda Jablonski or *PIW*.[22]

Sometime later, after Exxon reinstated its subscriptions, Chairman Ken Jamieson saw Wanda at an industry gathering and apologized to her, blaming the cancellation on "a colleague."

"Why, Ken," she replied. "I thought you were the chairman."[23]

PRIVATE NETWORKING

Despite their forewarning from Wanda and *PIW*, the majors had a hard time adjusting to the new reality that they no longer had unequivocal control over Middle East oil. During those turbulent days, Wanda listened and advised but did not significantly influence developments. Fifteen years earlier, she could provoke policy changes, but by the mid-1970s, international politics had overwhelmed her world. Despite her flurry of articles in late 1972, she left most of the weekly reporting to her staff. Instead, she mined her network quietly, focusing on a few key players of particular influence.

For the majors, the news only got worse. Emboldened by OPEC's growing leverage and worried about Israel's continued presence in the Palestinian territories, Saudi Arabia sent signals that Washington had better change its pro-Israel policies—signals that came through loud and clear in the pages of *PIW*. Indeed, those forewarnings were as good if not better than what the White House was getting in its daily intelligence briefings, according to a Senate committee that investigated the 1973 oil crisis. Not only had *PIW* reported those signals, as had other publications, but it gave evidence of tightening oil supplies and a tanker shortage. It also quoted senior oilmen acknowledging their weakness in the face of these new threats: Shell's Frank McFadzean conceded that the OPEC countries could "dictate—indeed they have dictated—the terms on which they'll trade their oil." In the same April issue, *PIW* also reported a warning from Yamani, which the publication said was the first clear indication that Saudi Arabia might feel compelled to use "oil as a political weapon."[24]

The surprise Arab attack on Israel in October 1973 was soon followed by what should not have been a surprise. Led by Saudi Arabia, the Arab producing states unsheathed the oil weapon, and this time it hit the mark. On October 16, five Arab Gulf states and Iran announced that for the first time they were unilaterally setting the price for their crude oil: it would be $5 a barrel, an increase of at least $2 a barrel. The next day, the Arab states announced significant production cuts across the board, with threats of steeper reductions until Israel withdrew from the occupied territories. Then, on October 20—the day after the U.S. government proposed a new military aid package for Israel—Saudi Arabia imposed a complete oil embargo on the United States, a move quickly followed by other Arab countries. The majors were in shock; *PIW* reported that they were "more deeply gloomy over the whole outlook than at any time in oil history." The embargo lasted five months. By March 1974 the price of Middle East crude oil had jumped to nearly $10 a barrel—a tripling in five months.[25]

Wanda was one of the few people who personally knew many of the key players in this crisis: the oil-club leaders and the national oil ministers. But as the conflict turned into a political and military showdown between nation-states, she watched from the sidelines. Nonetheless, she was much sought after for her private counsel. Although she did not like many Washington officials, she became an important source of insight and advice for James Critchfield, the CIA's chief intelligence officer on energy and a White House advisor in 1973. Aside from all the intelligence he traded with her, he also enjoyed her company and even dated her briefly. "She was the kind of woman," he said, "who could really talk with men."[26]

Among the oil chiefs she saw regularly, Mobil president Bill Tavoulareas was probably her closest contact and a personal friend. "They often talked about Yamani, whom they both knew well," recalled Mobil chairman Rawleigh Warner. But sometimes *PIW* articles strained their friendship. When markets reporter Marshall Thomas dug up sensitive information about Aramco reselling Saudi crude against Saudi wishes, for instance, he found plenty of evidence that Mobil was the worst offender. Early on the Monday the story was published, Thomas's phone rang. "You son-of-a-bitch," Wanda began. "The Greek just woke me up out of sound sleep screaming with rage!" Wanda had never killed any of his stories—and never did—but Thomas knew this one would upset Tavoulareas. Thomas had called Wanda about his scoop the week before, following her procedure on how to handle any article on Saudi Arabia: read it to Wanda in advance. This time, when she heard the story, she told Thomas in a clipped voice, "Check your numbers." He had, of course, but he checked again. His numbers were right—a little too right for Mobil.[27]

Of all her professional friends, the most important to Wanda was Yamani. According to two company chairmen, David Barran and Raw-leigh Warner, she, more than anyone else in the 1960s, had promoted the young Saudi minister and his pragmatic approach to oil national-ism. Wanda did not carry messages to the majors for him, the way she did for Tariki, but "talking with her was a way of expressing my views," he said. Yamani was quite cautious in speaking with reporters (he was so suspicious of journalists that he would often tape their interviews for himself), but his discussions with Wanda—whether at her home in New York or his homes in Saudi Arabia, Lebanon, Britain, and, later, Sar-dinia—were free-wheeling and always off the record, except for the formal interviews he gave her almost annually for nearly two decades. When he became a media sensation in the 1970s, Wanda did not change the way she behaved with him. She remained "very demand-ing," he said, "very imposing—and extremely honest." Most signifi-cantly, she made him think: "The way she asked questions would open windows for me."[28]

Did she make him more practical, as some oil-club leaders asserted? "Yes, in a way she did. She may have speeded up the process" of OPEC's rise to power, Yamani said in an interview. "The oil companies *did* want their business to remain a mystery. It was a real mystery in the 1950s and 1960s, and they hated to have anyone learn about it and try to join in the marketing business. That's why we had so many problems with negoti-ating participation." Her information was valuable "to both sides and gave her a lot of power," he said. Once the struggle over oil turned into a political and military conflict in 1973, he sought her views on occasion, but she was no longer a pivotal player. Yet when he was fired in 1986, as noted earlier, she came to his rescue with her final scoop about why the king had dismissed him.[29]

Wanda remained close to Yamani until the end of her life. He would call her frequently, often once a week. She was also a mentor to Yamani's daughter, Mai, who got to know Wanda in the 1970s while studying for a Ph.D. at Oxford. "I am fascinated by women who broke barriers, who had to be hard and pushy, who had to be really tough to survive in a world where men were the masters," Mai Yamani later said. "She made them treat her as an equal."[30]

LONELY AT THE TOP

With the 1973 Arab embargo, the oil world was turned on its head overnight. Wanda had hoped for a gradual transition to a more equitable balance, with the market—rather than a club or cartel—adjusting prices and production. However, as an entrepreneur in the news business, she succeeded spectacularly. Turbulent times made for lively news coverage. Her already authoritative publication became indispensable to an even broader audience. Suddenly oil was hot news, and the mainstream press called Wanda and *PIW* for comments and guidance.

Over the next few years, Wanda herself became the subject of numerous press reports, although she was uncomfortable with the attention. "Wanda Jablonski sits on top of the world—the oil world," began a 1973 feature article in the *Christian Science Monitor*. The Arab oil embargo "won't be lifted," she said, "until we produce concrete action on Israel." A resolution of the Arab-Israeli crisis, with Israel returning the land it had occupied since 1967, was critical. "What do we want?" she asked, with remarkable prescience. "A Soviet Union–U.S. showdown in the Middle East? Or do we want to send troops in to get the Arab oil? Then we'll have a showdown and no oil." Given the decline in U.S. reserves, Americans needed to cut consumption through strong conservation measures, stimulate output by decontrolling oil prices, and develop alternate sources of energy "so we won't be so dependent on foreign oil twenty years from now."

Notably, her personal views on the crisis did not affect *PIW*'s reporting. "She was a total pragmatist when it came to covering the Middle East and the markets," said Thomas. "She didn't want *PIW* to take any side on the Arab-Israeli fight."[31]

In a laudatory 1977 piece in the *New York Times* (and published in the women's section), Nan Robertson wrote that Wanda Jablonski "may know more about the people who have oil and the people who want it than just about anybody in the world."[32] Those words were echoed on the *CBS Evening News* later that year, when Steve Young reported: "Odds are you've never heard of Wanda Jablonski, but just ask oil men like the president of Mobil, or Sheikh Ahmed Zaki Yamani, oil minister of Saudi Arabia. They'll tell you that she may know more about oil than anybody." While Young explained what Wanda did, the camera followed her as she talked with people at a United Nations cocktail party. "Let me jot that down," she said. "Somebody give me a pencil." Wanda and her staff were known for letting "the chips fall where they may," Young said, even though it occasionally meant boycotts by Big Oil and Arab potentates for printing embarrassing articles. "I want the finest publication in the world, and I'm very demanding, I'm absolutely impossible, I'm temperamental, and I want everything yesterday, and I want it all checked out, and I want it 1,000% accurate," Wanda said. "Knowing I'm impossible to work for, I've gotten myself, in due course over the years, an editor who has got all the qualities I want and who is just superb." An oil executive at the party then said, "I think *PIW* is more accurate than the press in general."[33]

Though "few fields are as male-dominated as oil," CBS's Young continued, Wanda Jablonski went to the Middle East in the 1950s, "back when a woman oil expert traveling alone in an Arab country was considered an astonishing phenomenon." As early as 1955, she had warned that the United States was becoming too dependent on Arab oil, and that was still true. Looking straight into the camera, Wanda said the leaders of OPEC agreed with her: "They say your country's crazy. You're making yourself more and more dependent on us, and then kicking us around." The camera then followed her slipping through a cluster of people as Young concluded, saying, "Tiny, tough Wanda Jablonski melts into another oil crowd to see what she can learn for next week's edition."[34]

Although CBS's description of Wanda's involvement in *PIW* at that time was exaggerated, Macris and her reporters were proud of her accomplishments—and resentful of her high-handed ways. At one point in the 1970s, when she admitted being bored, she started coming to the office regularly on Thursdays, the day the paper went to press. She "helped" by marking up copy with her fat editor's pencil. "Wanda would pace back and forth with a cigarette, not necessarily lit," Lou Moscatello, a staff reporter, remembered. "She was like a caged cat, thinking of what to do next, who to call or what to write. She was just naturally wired." But she made the staff nervous, and she sometimes treated Macris in a demeaning way. John Lichtblau, the well-known oil consultant, remembered being appalled when Wanda once "humiliated" Macris in front of him. Finally, Macris told her—once again—that two cooks in the kitchen was one too many. So Wanda got out.[35]

Still, the founder and publisher of *Petroleum Intelligence Weekly* remained its guiding force. Despite Wanda's quarrelsome nature, she attracted and retained a talented group of reporters and editors. Having started the company without family money or a wealthy backer, she became a true entrepreneur, building the publication on the strength of her own reputation. As the *Wall Street Journal*'s James Tanner said, "She changed the face of oil journalism" by setting a higher standard. She also created a new kind of newsletter: a must-read trade publication for a particular industry, not the general public. Her goal was to get specialized business intelligence to an elite audience through sheer market sleuthing. Before she launched *PIW*, this kind of investigative and analytical business journalism was rare. Even as the crises of the 1970s led to more sophisticated oil reporting in national dailies and the business press, *PIW* maintained and broadened its audience. "The general press, especially television, can't handle nuance," said *PIW*'s Moscatello, "whereas nuance was the name of the game at *PIW*."[36]

THE WALKOUT

In the early 1980s, however, Wanda began to have serious trouble with *PIW*, trouble that would nearly lead to its collapse. One set of problems came from Wanda's reluctance to adapt quickly to changes in the oil market, in computer technology, and even in business reporting—all of critical importance to *PIW*. During the 1970s, influence over crude-oil prices shifted from the oil-club leaders to the OPEC countries, and then, by the early 1980s, to oil traders operating in the global cargo market. Once the U.S. government abolished its remaining price and import controls in 1981, commodity trading in oil futures took off in London and New York. By bidding on the marginal barrel based on the actual or perceived supply-and-demand conditions, futures exchange traders set prices. In particular, the major oil companies encouraged oil trading centered on sources outside OPEC's control, such as North Sea Brent and U.S. West Texas Intermediate crude. Increasingly, these prices, rather than the reference prices set by OPEC, were accepted as true indicators of oil's value. As a result, the market became preeminent.[37]

By 1975, *PIW* was providing such detailed coverage of changes in prices, production rates, and freight rates that anyone interested in international oil studied it "like a baseball fan poring over batting averages," as one report said. Although Thomas's coverage of prices on the flourishing "spot" (short-term) market became quite influential, Wanda was unwilling to provide the money and staff needed to develop timely market coverage.[38] As the market became more liquid, the oil industry needed daily or even real-time information—something that *PIW* in its traditional format could not provide. After repeated delays, she eventually allowed Thomas to start a daily electronic oil-pricing service with other market specialists, but she soon sold *PIW*'s share in the joint venture to McGraw-Hill. "Big personalities in the oil business, not market pricing" remained her main interest, said *Petroleum Argus* publisher Jan Nasmyth. By restraining *PIW*'s market coverage, she kept the staff from adapting quickly to the market needs, which allowed other publications to fill the breach.[39]

But the most significant threat to *PIW*'s future came from Wanda's deteriorating relationship with her staff and her inability to deal with her succession. After Georgia Macris retired in 1983, Wanda tried to persuade Marshall Thomas to become editor—he was her "star performer," she later said—but he shied away from the job. Though well liked by the staff, he was a market sleuth and did not have the patience to be a desk editor. Wanda then tried a triumvirate of editors: Thomas; Ken Miller, who had returned to New York after reporting for nearly two decades abroad; and Tom Wallin, who had managed *PIW*'s London office for several years. But without clear lines of authority, organization was difficult. Wanda "was juggling these people like puppets," Macris said. "She had an almost feudal mentality—this is my property, so I can do as I please. She could be so generous sometimes, and then be so petty and greedy." Wanda's arbitrary management style did not work well with a small staff.[40]

What Wanda cared most deeply about by the 1980s, said Macris, was that *PIW* continue after her death. Yet every effort she had made to find and groom a successor had failed. She found fault with every longtime staff member, no matter how talented, and she—as well as Macris, sometimes—undermined outside prospects, including *New York Times* reporter Anthony Parisi. A London-based energy-markets specialist, Patrick Heren, said he twice turned down offers from Wanda because he had heard too many stories about her difficult temperament. In 1985, Wanda hired John Treat, first president of the New York Mercantile Exchange, to take over as executive publisher, but she soon forced him out because of staff objections. Although Wanda had arranged to leave the newsletter to a board of trustees with an ample trust fund, she kept changing the terms and the trustees, only further unsettling her longtime staff.[41]

Morale was so bad by the mid-1980s that several reporters secretly drew up a business plan to attract potential investors and start a competing publication. In their draft plan, they recognized how tough the prospects were: "WJ is a shrewd and ruthless negotiator. It is imperative that she be allowed no time to mount a counter attack." But the effort foundered amid concerns about the legality of the challenge, the financial hurdles, and the impossibility of attracting investors without Wanda finding out.[42]

Two things finally triggered Wanda's decision to sell her newsletter. In March 1988, Marshall Thomas told her he would start his own consulting business by mid-June. Shortly thereafter she had a dispute with William Liscom, *PIW*'s acting business manager, who also said he would leave by June. The next day Wanda called the chairman of pub-

Wanda with her longtime secretary, Agatha Sangiorgio ("Agnes Stewart"), and Georgia Macris at Macris's retirement party on the terrace of Wanda's penthouse, New York, 1983.

lishing giant International Thomson, who had tried several times to buy *PIW*. He said he was willing to offer $15 million, but Liscom undermined the deal, several staff members later charged, by telling Thomson's team that *PIW* was in terrible shape and was about to lose key staff members, even though only Thomas was leaving. Most of the staff thought Thomson was an attractive buyer and were upset when it suddenly withdrew its offer. Wanda then decided that the best alternative was an offer from a Washington-based consulting firm, Petroleum Finance, run by Edward Morse and Robin West. Morse assured her they could set up a "Chinese wall" between the two operations—consulting and reporting—to ensure *PIW*'s independence and integrity.[43]

But the staff objected. Wanda's management in recent years had left most of them quite upset—even bitter—over a canceled pension plan (her promise of an alternative plan never materialized) and a cutback in full medical coverage. She had also caught senior editors off-guard with her sudden decision to sell the publication. For some time, Liscom had been urging the staff to force Wanda's hand by walking out so she would give them partial ownership and a genuine management role. When Thomson backed out in early May, Liscom persuaded most of the staff, including Avati in Paris and several in *PIW*'s London office, to sign employment contracts with him to use as leverage with Wanda. Many of them did not think they would actually work for Liscom. Wanda was expected to cave in to a walkout by the entire staff. "There was no real plan for what to do if she didn't give in," said copy editor Guy DiRoma. When Liscom called Wanda on Memorial Day to offer the staff's employment contracts for $4 million, she turned him down. She would not cede to "blackmail," she said. The next day, Morse talked with the staff at her request to explain Petroleum Finance's offer, but Liscom persuaded most of them to walk out; the only main exception was Tom Wallin, executive editor in New York.[44]

It was a "bizarre chain of events," the *Wall Street Journal*'s Tanner reported the following week. The mass defections and start-up of a new oil information service, called *OPEC Listener*, "is livening up an otherwise dull period." *Petroleum Intelligence Weekly*, Tanner added, had been "a publishing gem" largely because of its "staff of top energy journalists."[45]

Virtually alone, Wallin and Wanda put out the next few issues before finding some replacements. Wanda pitched in on everything from writing headlines to answering the phone. "Sure, that's my dog," she told callers if they heard a bark. "We need all the help we can get." *PIW* got additional support from Petroleum Finance, once Wanda had tentatively accepted its offer (one Petroleum Finance staffer had already been helping out), and succeeded in hiring some knowledgeable reporters—Amy Myers Jaffe in New York and Patrick Heren in London. *PIW*'s ability to rapidly rebuild its staff undermined prospects for the ex-staffers' new publication, as did the multimillion-dollar lawsuit that Wanda had filed against them. She quickly got a court order to block a rival newsletter, but this did not affect Liscom's *OPEC Listener* because it was a facsimile news service. Within six months the restrictions were reduced, but the lawsuit endured until Liscom and the ex-staffers eventually settled out of court.[46]

This was the biggest battle of her life, she told friends. She conceded that she bore "some responsibility" for the walkout and needed others to help her ensure the future of what she called her "baby." On July 7 she sold her newsletter to Petroleum Finance for a sum that was undisclosed but was less than $12 million. At the time, about three thousand official subscribers were paying $1,150 annually for the eight- to twelve-page weekly, and there were some additional subscribers to a monthly oil price report. For the new executive publisher, Ed Morse, the eighteen-month transition period with Wanda was tough: "She was—as anyone who worked closely with her knows—exceptionally blunt." She had "qualms over selling her baby to a virtual stranger." He, in turn, helped her "let go."

Ironically, even though she had a wide range of people on her staff, Wanda was reluctant to have a Jewish or an Arab editor or publisher "because of the political ramifications," Thomas said. But after consulting Arab confidants, she decided Morse would work out. In 1996, Raja Sidawi, owner of the Washington-based *Oil Daily* and member of a Syrian publishing family, bought *PIW* and its affiliated newsletters. It is now the flagship of a dozen publications—still leaders in the oil industry—for Sidawi's Energy Intelligence.[47]

Fortunately for Wanda, before she sold *PIW* she had already found another venue for her restless energy: the Strang Cancer Prevention Center in New York. When she had a benign lumpectomy in the early 1980s, she decided to help the director, Dr. Michael Osborne, develop a national high-risk breast-cancer registry, and in 1991 she donated $3 million to the center. Osborne became, she later said, the "son she never had—and never wanted."[48]

In 1988, she also visited relatives in Bratislava for the first time since childhood. Her cousins Vladimir and Elena Krcmery, both medical doctors, received her several times over the next three years and took her to see her birthplace, Trnava. Her visit in November 1989 was during the "dangerous days" of Czechoslovakia's so-called Velvet Revolution, Vladimir Krcmery said. "She came from OPEC in Vienna by a personal taxi. [The] revolution was in full course, [the] streets were full with people and with police and army—and she had no fear," he wrote. "She brought money to our families in envelopes named by several delegates of OPEC." If she had been stopped by the police, as she had been the year before, she was ready to say that the money was for the OPEC delegates. Wanda delighted in this renewed contact with relatives and helped finance medical studies in the United States for the Krcmerys' daughter, Maria.[49]

Wanda's own medical problems, however, soon emerged as quite serious. A chain-smoker since her youth, she had tried numerous times to quit, but could not. Osborne even allowed her to keep up her two-pack-a-day habit while volunteering at his cancer center, though only in a private office. Tests in late 1991 showed that she had lung cancer. In December, she called her Slovak cousins, Vladimir and Elena, to tell them she was going to have surgery but would complete arrangements for their daughter's U.S. studies beforehand. Admitted January 15 to New York Hospital, she developed complications after her operation and died there from heart failure on January 28, 1992.[50]

The Strang Center was her primary beneficiary. Although she left small sums to her relatives in Slovakia and her longtime secretary, Agatha Sangiorgio, Strang received $27 million—its largest donation ever. PIW sponsored a public gathering to remember her legacy as the newsletter's founder, attended by former colleagues and oil industry contacts, and issued a special supplement about her life's work. But just eight people went to a private memorial ceremony in April: her cousin Vladimir and his daughter, Maria; Agatha Sangiorgio; Dr. Osborne; her accountant, Ted Schnall, and his wife; her lawyer and executor, Michael Saltser; and one fairly new friend, Denise Regan. At a Long Island beach, they put some of her ashes in the water at sunset, along with eight red roses, one by one.[51]

THE LAST SCOOP

As her colleagues, former colleagues, professional contacts, and personal friends took stock of Wanda's life at PIW's tribute in 1992, some of them still puzzled over her last scoop: the true story behind the fall of her friend Zaki Yamani in 1986. With the Gulf War of 1991, oil prices had once again shot upward, and concern about access to sufficient supplies returned to the front pages. For old-timers, this was just another upswing in the roller-coaster ride of supply and demand. But the price collapse in 1986—OPEC's own "oil shock"—had changed everyone's perspective. Yamani had been right: OPEC's high prices led to substantial conservation, greater industrial efficiencies, new oil supplies from non-OPEC sources, and improved technology from the drill bit to the gasoline pump. They also led to a worldwide economic recession, further reducing demand.

The initial consequences were most acute for Saudi Arabia. With other OPEC members unwilling to restrict production, Saudi Arabia took on the responsibility of "swing" producer. As supply needs rose with the start of the Iran-Iraq war in 1980, it raised production but soon had to cut back—and then cut back dramatically. By 1985, Saudi Arabia was down to barely one-fifth of its peak level of 10 million barrels a day. For a country in the midst of a huge development program, that meant a third year of serious budget problems. As the worldwide surplus grew, OPEC countries had agreed to individual country production quotas to help pare down the glut, but regularly cheated on these quotas and undercut official prices through unofficial discounts. So the Saudis decided to increase their own production in September 1985.

Reversing his formerly staunch support of fixed prices, Yamani began to sell crude to the majors at prices linked to refined products. In effect, Saudi Arabia declared war against all other producers, including OPEC members, as a way to regain market share. The effect was almost immediate. Prices fell precipitously—from more than $31 a barrel in the fall of 1985 to barely $12 in six months.[52] The price of oil hit a new low in mid-July 1986, the *Wall Street Journal* reported, because of a story in *PIW*, which King Fahd immediately but unsuccessfully sought to counteract. On July 14, *PIW* reported that OPEC's output, instead of declining as it should have from the organization's latest agreement, had reached a "stratospheric" 19.5 million barrels per day, the highest level since 1982. That sent market prices tumbling, dipping to less than $9 a barrel in Europe and less than $11 on the New York futures market. The next day, King Fahd announced that Yamani would "take steps to seek stability in world oil markets," but prices barely budged. Many oil traders—those whose buying and selling activities were now setting oil prices on the spot and futures markets—remained skeptical.[53]

But that price collapse put Yamani's job in jeopardy. His relations with King Fahd were strained, and a particularly thorny issue was the bartering of oil. Yamani had been critical of this practice because other OPEC countries were, in effect, cheating on their quota and their price commitment by not including bartered barrels, forcing Saudi Arabia as swing producer to cut its production still further to rein in the glut. But Yamani was overruled in 1984, when the royal family insisted on its own $1.3 billion barter deal (34 million barrels for Boeing jets), and a $2 billion oil-for-jets deal in 1985.[54]

Yamani's problems multiplied in 1986 as the king insisted that he pursue a new double-pronged policy: a quick return to fixed prices set at $18 a barrel, and higher Saudi production. At OPEC's August 1986 meeting, Yamani explained the new policy, but observers noted that he did so without conviction. When OPEC met again in October, Yamani's tepid support for this new policy was obvious. Nonetheless, when Yamani was fired a week later, the oil world was stunned, with front-page articles speculating about why OPEC's leader for nearly a quarter century had been dismissed so callously. Most attributed it to a dispute with the royal family, though some mentioned a secret discount Yamani was offering select customers. But even two weeks after his firing, the *Economist* observed that the kingdom's decisions "have grown curiouser and curiouser." Why, indeed, had the world's most famous oil minister been ousted?[55]

For many years, Yamani refrained from commenting on the circumstances of his dismissal, despite repeated questioning. But when approached for an interview about Wanda, he decided to set out his perspective because of the distortions he had seen in numerous accounts, and because he wanted to make clear that it was Wanda who had rescued him, in effect, from house arrest.

Yamani had been expecting his dismissal for some time, he said, but it was still difficult. On October 29, the day he heard the announcement on late-night television, he received a phone call at his home in Riyadh from his close friend and fellow oil minister Ali Khalifa al-Sabah of Kuwait, who warned him about a conversation that had just taken place. Earlier that day, Khalifa recently recalled, Kuwait's ruler, Jaber al-Sabah, had taken a call "from King Fahd indicating that Saudi Arabia and Kuwait should [discuss] their oil policies." Since they were to meet three days later in Abu Dhabi, Jaber suggested that Khalifa and Yamani draw up a policy paper for the two rulers to discuss. King Fahd agreed but suggested that Khalifa be joined instead by Hisham Nazer, Saudi Arabia's minister of planning. The emir called Khalifa immediately to inform him, because he knew Khalifa was about to leave for his seaside estate. "We concluded," Khalifa said, that Yamani "was about to be relieved of his duties." The Kuwaiti oil minister was shaken by the news. "I thought that I better give myself a chance to collect my thoughts on the one-hour drive to my chalet," he recalled. When he arrived, "I called Sheikh Zaki and strongly hinted of the essence of the conversation, so he had some ten hours early warning from me."[56]

Right after Yamani's late-night ouster, "there were some real noises in the media about why I left," Yamani said. "This annoyed certain members of the royal family. They wanted at any price to ruin my image." Despite intense efforts to find evidence pointing to possible corruption by searching bank and other records, "they found nothing." Nonetheless, they started a "smear campaign" against him, he said. They planted false information about a Saudi "committee to investigate Yamani's corruption" in an Egyptian publication, which they had widely distributed. This "nonsense" was picked up by some in the Western press. Yamani was so angry that he decided "to be open and write articles and counterattack what they were saying because I have so much to reveal." Since they were trying to hurt him, "I could really hurt them." He was also quite upset because the royal family had taken away his passport and put him under virtual house arrest at his home in Jeddah. "Every phone call I made was being recorded and sent to the king. I had the intelligence [service] outside my house" watching all visitors. And he had hundreds of visitors: "It was a way of showing solidarity."[57]

Wanda "saved me from all this trouble," Yamani said. Four weeks after his ouster, *Petroleum Intelligence Weekly* published the "inside story" of what had happened: "The real conflict" between Yamani and the Saudi king "ran considerably deeper than speculations that have appeared so far in the world press," *PIW* reported. The "final showdown" came when Yamani disputed King Fahd's order that Saudi Arabia should secretly violate its OPEC production ceiling by not reporting the oil it was using for barter deals. "The King indicated that barter oil could be handled outside the Saudi OPEC allowable, and need not necessarily be mentioned to OPEC," *PIW* said, attributing the information to "a highly reliable and responsible source close to the situation." In effect, the king wanted him to "cheat" on its OPEC production quota. Yamani argued instead "that Saudi Arabia had 'a good record' in abiding by OPEC regulations and production reporting" and that, in any case, the excess production would be impossible to hide.[58]

This insubordination was the latest of a series of confrontations. During the October OPEC meeting, Yamani refused the king's instructions to raise Saudi Arabia's production quota to no less than five million barrels a day, and preferably six million, as well as to raise OPEC's price to no less than $18 a barrel, and preferably $20. The king had apparently been assured by other advisors that both goals were achievable. Yamani, according to *PIW*, told the king that the glut made it "impossible to achieve both goals simultaneously," and the king's "two-pronged policy would amount to 'economic suicide' for Saudi Arabia and [he] refused to accept responsibility for it."

What's more, the barter deals were far larger than anyone had previously known. Aside from the two oil-for-aircraft deals with British and U.S. companies that had been reported, there were two more barter deals in the works: one for construction of a new refinery on the Red Sea and another for construction of an underground products storage facility. These deals amounted to two million barrels a day—more than half of Saudi Arabia's production quota, *PIW* said. "Yamani opposed all such barter deals" because the oil would be "dumped on the world market." When the king concluded that his oil minister "was deliberately acting against his wishes, and consistently throwing up obstacles," the newsletter stated, Yamani was "abruptly discharged."[59]

Yamani, in his interview, confirmed this story. He insisted that he did not know how Wanda got the information, or who the mysterious caller was who gave it to her. What he did know was that very few people had this information, and that publishing it "was really something very important." Because of what it revealed, because it was widely read and discussed in Saudi Arabia, "the royal family members who had started the campaign against me got together with other family members and decided to shut their mouths. To continue would open up the possibility of more," Yamani explained. "'If Wanda Jablonski can do this,' they said to themselves, 'others might do the same.' So they stopped. She stopped the royal family just like that."[60]

And once again, with one last scoop in the twilight of her career, Wanda Jablonski surprised the oil world. "It was a gutsy story," said the *Wall Street Journal*'s James Tanner. "The information on the barter deals was all new to us—that was the real news in her story. The Saudis were very upset by it." What struck Tanner as significant was that Wanda was willing to take such a risk—the Saudi royal family could damage any publication, even one as famous as *PIW*. Kuwait's former oil minister, Ali Khalifa, agreed that Wanda had the story right, but he said the risk she took was not so significant because by then the oil industry was fairly transparent. What's more, he said, "no country or oil company can really punish a widely read publication." It had become too critical to the industry.[61]

Three years later, when Wanda stopped off in Vienna in 1989 to see old friends at what would be her final OPEC meeting, Tanner watched Yamani's successor, Hisham Nazer—"someone who I know was very upset with her"—break loose from a bevy of reporters to shake hands with her. "That gracious greeting," Tanner commented, "said a lot."[62]

A PRISM

During the 1980s, as oil prices crippled the world economy and then collapsed, Wanda would fly to Britain once a year to attend Professor Robert Mabro's Oxford Energy Seminar, where the most select leaders in the oil world—executives and government ministers alike—came to network. Still vigorous in her sixties, she had outlasted and even outlived several generations of company chairmen. Perfectly coiffed, dressed in fashionable pant suits, she was often the only woman in the room. She would sit through long sessions, Mabro remembered, "almost motionless, silent, inscrutable." Between sessions, she would occasionally whisper to him unflattering remarks about some of the eminent speakers. "She knew them all, their history and their character, their weaknesses and idiosyncrasies. She was cynical about men and events, but her cynicism was mixed with a tinge of amusement," he recalled. "Behind the façade, and despite a tremendous act of concealment, there was a glimpse of a soul."

Mabro tried to persuade her to write, or even dictate, her memoirs. Had she not "killed" thousands of stories for which other journalists would have given the earth? "She knew so much, had witnessed so many events." The personal stories she told were "no doubt embellished, but rich in insights." But she turned him down. "I am a journalist," she replied. "The past is for historians."[63]

Those tales turned out to be, for the most part, remarkably accurate. Indeed, Wanda's extraordinary life bears witness to the power of information: how it can clarify issues, focus minds, and force decisions. At a time when tensions over the Middle East once again dominate the world's news, when images of American soldiers in Iraq are deeply imprinted on the world's consciousness, Wanda Jablonski's life is a prism through which to look at the formative years of American involvement in that region. It shows how profound cultural, economic, and geopolitical misunderstandings have marked the relationship between the United States and the Middle East. Fifty years ago, donkeys delivered water in goatskin sacks in Qatar, American oil imports from the Middle East were but a trickle, and a female oil reporter appeared so odd that, as she wrote from Saudi Arabia on Christmas Eve 1956, she felt as though she had two heads. Fifty years ago, Wanda also told an international affairs gathering that Americans needed to "learn to think how the Arabs think, not how *we* think they should think." And more than thirty years ago, she warned that the United States had to cut its oil consumption because if it sent in troops to secure Arab oil, "we'll have a showdown and no oil."[64]

Though Wanda did not see herself as a feminist, this female reporter—who used her initials to hide her sex for a decade, but then became famous simply as "Wanda"—defied expectations and broke a path for other women to follow. She came to define a new style of investigative business journalism, at once more attuned to differing perspectives and more sensitive to the long-term implications of daily news. By breaking the mold, by questioning and probing and continually refining her own perspective, she helped change seemingly fixed parameters. She got people thinking in new ways.

In an industry in which secrecy reigned supreme, Wanda repeatedly demonstrated that personal access, exclusives, and getting the scoop could influence people and the course of events. Sometimes she affected specific decisions—Rathbone's price cut or the shah's unexpected statement on OPEC—and sometimes her reporting influenced developments over time. Recognizing the significance of oil nationalism, she drew attention to the potential power of OPEC's founders, Juan Pablo Perez Alfonso and Abdullah Tariki, while at the same time educating these ministers in the ways of the oil industry. By exposing and explaining many mysteries of the oil trade, by identifying the pressure points that made the major oil companies vulnerable, she helped accelerate the producing countries' efforts to gain greater control from Big Oil. In effect, she helped dispel the aura of invincibility that had cloaked the Seven Sisters' club for so long. No other business reporter in the second half of the twentieth century had as much influence over an industry as Wanda Jablonski had over international oil.[65] She mastered that world and left her mark.

Acknowledgments

Writing a woman's life has been a long, complicated, but immensely enriching experience for me. When I started working as a reporter in New York in the 1980s, women were making significant strides in the workforce. Many subtle and not-so-subtle barriers to advancement remained, but it was no longer the hostile environment that female professionals faced in the post–World War II years. Women writers articulating new kinds of narratives inspired me to undertake and finally complete Wanda's story. We need more biographies of compelling women, women who distinguished themselves in spite of their flaws, women who questioned and learned and changed the way people thought.

I would like to acknowledge people who have been instrumental in shepherding me through my own journey. Central to my work as a historian is John Morton Blum, the distinguished professor emeritus at Yale University, who not only gave me superb training as a scholar, but persuaded me to return to Yale to write my dissertation about Wanda and her world, grinned at what he called my "journalese," and sustained me with his friendship. My special thanks also to my two dissertation advisors, Diane Kunz and Cynthia Russett, as well as to Florence Thomas; and to two inspiring professors at Vassar College, David Schalk and Robert Fortna. I am also indebted to Yale for two critical fellowships.

This book would not have happened without Daniel Yergin, chairman of Cambridge Energy Research Associates. His advice and enthusiasm helped me at every stage: he prodded me to start, read the entire dissertation during a holiday, delighted in my discoveries, persuaded me that I could turn a partially completed biography into a story with wide appeal, edited various drafts, and graciously wrote the foreword. He is the very definition of "mentor."

I am most appreciative of the work of my agent, Lisa Adams of the Garamond Agency. She more than deserves her excellent reputation. Her wise advice, marketing savvy, and skilled editing have greatly improved the book. I would also like to thank the entire staff of Beacon Press, especially my editors, Christine Cipriani and Gayatri Patnaik, for enthusiastically embracing this project. In addition, I would like to thank Joanne Wyckoff, Emily Herring Wilson, and Peter Hellman for their judicious advice on the publishing world.

Because Wanda left only a small collection of private letters and files, oral history was essential to flesh out her story. I am deeply grateful to more than one hundred persons who talked with me—some at great length—about their recollections of Wanda; they are listed at the end of these acknowledgments. I also particularly appreciate the exclusive access to Wanda's papers that I received from Edward Morse, previous publisher of *Petroleum Intelligence Weekly*; Raja Sadawi, chairman of Energy Intelligence; and Thomas Wallin, president of EI. Tom's support has been critical over many years and in many ways. I especially thank Ambassador Brandon Grove, Jr., Onnic Marashian, Marshall Thomas, and Dan Yergin for their willingness to share their private papers, and Sir Denis Wright, for sharing his unpublished memoirs. My thanks as well to the staffs of the Library of Congress, the National Archives (United States), the British Public Record Office, the Wake Forest University Library, the Georgetown University Library, the Amherst College Library, and the New York Public Library.

Many personal friends helped me along the way. Biographer Penelope Niven coaxed me into launching this effort and my "BioBrio" writers' group at Wake Forest University—Margaret Supplee Smith, Michele Gillespie, and our dynamic leader, Emily Herring Wilson—would not let me stop. As I traveled to complete my interviewing, I was graciously welcomed by Sophie Nicolas, Joan O'Connor and Bob Jones, Wendy Phillips Kahn, Anne and Yves Girault, Sheila and Ed Becker, Anne Easterbrook, Cecilia Rubino and Peter Lucas, Helen Rubino-Turco and Paul Turco, as well as my parents. My special thanks to Elisabeth Hausmaninger for her help with German translation; and to

Jon Gates, Off The Record's director of research, and Lisa Warren, my OTR editor, for supporting my efforts to complete this book. I benefited greatly from the comments of friends and fellow reporters who read all or portions of this book, including Gale Sigal, Gail Fisher, Annette Porter, Pat Maida, Helen Avati, Tom Wallin, Sarah Miller, Youssef Ibrahim, Lou Moscatello, Marshall Thomas, Onnic Marashian, John Buckley, Napier Collyns, David Knapp, Morry Adelman, Peter Green-leaf, Adrian Binks, and Jan Collins, as well as Patrick Heren, who, at a critical moment, read and critiqued the entire manuscript within days. I also owe a special debt of gratitude to several friends deeply involved in this book since its inception: Diana Greene, a gifted writer who kept me walking through so much; Sarah Watts, a masterful historian who, with probing questions and critical insights, helped me from start to finish; and Diane Munro, a tenacious reporter and editor, who improved the book immeasurably with her deep knowledge of the industry and the main characters. She read draft after draft after draft and would not let me give up. My friends' critical readings much improved the text, but any errors are, of course, my own.

I would also like to thank all the members of my extended Rubino clan, each one so treasured, and especially my sisters, Margaret, Cecilia, and Helen, for nurturing me throughout, and my daughter Caroline for her extensive help with the manuscript. Most of all I thank my mother and my father, Nancy and Michael Rubino, for their steadfast support and abiding love. I am truly blessed. My father has devoted countless hours to this project, encouraging my every step, always eager to read, edit, network—anything I asked.

I dedicate this book to my husband, Dick Schneider, my wise editor, my refuge and sustainer, my wellspring of love. And I dedicate it to our daughters, Laurian, Caroline, and Elena—each one, pure joy; each one, a reflection of their devoted grandmother, Joan Laurian Brun-dage, whose ebullient spirit continues to live on through them. Her de-termination to break through barriers against women professionals in medicine—and still embrace life as it is—inspires all of us who knew and loved her.

Oral history was a critical part of my research. I greatly appreciate the willingness of more than one hundred persons, including several anonymous sources, to share with me their recollections and assessment of Wanda (affiliation, profession, or title dating from the time of the interview):

M. A. Adelman, professor emeritus, Massachusetts Institute of
    Technology
James Akins, former United States ambassador
Michael Ameen, former vice president, government relations, Aramco
Helen Avati, oil reporter (former Paris bureau chief, *Petroleum
    Intelligence Weekly*)
Sir David Barran, former managing director, Royal Dutch/Shell
Rostam "Mike" Bayandor, former executive, National Iranian Oil Co.
James Bill, professor, College of William & Mary
John Bishop, staff official, European Commission (former reporter,
    PIW)
William Bland, president, William F. Bland Co.
John Buckley, former editor, *PIW*
W. Jack Butler, former chairman, Mobil Saudi Arabia
Fadhil Chalabi, director, Centre for Global Energy Studies (former
    acting Secretary General, OPEC)
Margaret Clarke, former copyeditor, *PIW*
Napier Collyns, cofounder, Global Business Network (former
    Royal/Dutch Shell executive)
Harriet Costikyan, former circulation manager, *Petroleum Week*
James Critchfield, former national intelligence officer for energy,
    Central Intelligence Agency
Lois Critchfield, editor, Gulf Futures, Inc.
Guy DiRoma, former copyeditor, *PIW*
Sir Eric Drake, former chairman, British Petroleum
Andrew Ensor, former Middle East manager, Mobil
Paul Ensor, former general manager (Persian Gulf),
    Iraq Petroleum Co.
Manucher Farmanfarmaian, former director, NIOC
Dennis FitzPatrick, former executive, BP
Larry Goldstein, president, Petroleum Industry Research Foundation
Brandon Grove Jr., former United States ambassador

Susan Habachy, former reporter, *PIW*

J. E. Hartshorn, former senior reporter, *Economist* (former reporter, *PIW*)

Jack Hayes, former executive, Aramco

Barth Healey, staff editor, *New York Times* (former reporter, *PIW*)

Patrick Heren, chairman, Heren Energy (former editor, *World Gas Intelligence*)

Ann-Louise Hittle, senior director, Cambridge Energy Research Associates (former reporter, *PIW*)

James Hunt, former president, New York Society of Oil Analysts

Youssef Ibrahim, senior Middle East correspondent, *New York Times*

Amy Myers Jaffe, energy fellow, James A. Baker III Institute for Public Policy (former reporter, *PIW*)

John Jaqua, former partner, Sullivan & Cromwell

Joseph Johnston, former senior vice president, Aramco

Sheikh Ali Khalifa al-Sabah, former oil minister, Kuwait

Karol Krcmery, Wanda's cousin

Dr. Vladimir Krcmery, Wanda's cousin

Walter Levy, founder, W. J. Levy Associates

John Lichtblau, chairman, Petroleum Industry Research Foundation

William Lindenmuth, former senior executive, Mobil

John Loudon, former chairman, Royal Dutch/Shell

Beverly Lutz (formerly Parra), personal friend of Wanda's

Robert Mabro, director, Oxford Institute for Energy Studies

Howard Macdonald, former chairman, Dome Petroleum (former Shell executive)

Georgia Macris, former editor, *PIW*

Onnic Marashian, former editor, *Platt's Oilgram News*

Phebe Marr, fellow, Woodrow Wilson Center for International Scholars

Feizal Mazidi, former senior official, Ministry of Finance, Kuwait

Kenneth Miller, former editor, *PIW*

Leyna Miller

Sarah Miller, editor, *PIW*

Arthur Mills, former chauffeur, Mobil (London)

Parviz Mina, former director, International Affairs, NIOC

Edward Morse, publisher, *PIW*

Louis Moscatello, financial analyst and advisor (former reporter, *PIW*)

Henry Moses, former senior executive, Mobil

Alfred Munk, former senior executive, Amoco

Diane Munro, president and editor, Oil Daily Co. (former reporter, *PIW*)

Fathollah Naficy, former director, NIOC

Farrokh Najmabadi, senior consultant, World Bank (senior official, NIOC)

Jan Nasmyth, publisher, *Petroleum Argus*

Walter Newton, former managing director, Petroleum Economics Ltd.

Richard Nolte, former United States ambassador

Eugene Norris, former reporter, *PIW*

Dr. Michael Osborne, president, Strang Cancer Prevention Center

Alirio Parra, former minister of energy and mines, Venezuela

Francisco Parra, former secretary general, OPEC

John Pearson, former Latin America editor, *Business Week*

Ruth Pearson, reporter

Dame Edith Penrose, former professor, London School of Economics

William Randol, oil analyst, First Boston

Denise Regan, personal friend of Wanda's

Aaron Riches, former reporter, *Journal of Commerce*

Silvan Robinson, former president, Shell International Trading

Joe Roeber, president, Joe Roeber Associates

Michael Saltser, partner, Battle Fowler

Erich Schliemann, former director, Deutsche Erdoel

Herb Schmertz, former senior vice president, Mobil

Henry Schuler, former executive, Hunt Exploration and Mining Co.

Eleanor Schwartz, former reporter, *Journal of Commerce*

Ian Seymour, editor, *Middle East Economic Survey*

Eileen Shanahan, former reporter, *New York Times*

William Smith, former reporter, *New York Times*

Dillard Spriggs, oil industry consultant

C. B. Squire, former reporter, *Platt's Oilgram News*

Sister Mary Louise Sullivan, former president, Cabrini College

Jack Sunderland, former chairman, Aminoil

James Tanner, former reporter, *Wall Street Journal*

Fred Thackeray, consultant (former reporter, *Petroleum Economist*)

Marshall Thomas, consultant (former editor, *PIW*)

Lord Christopher Tugendhat, chairman, Abbey National (former energy editor, *Financial Times*)

Thomas Wallin, president, Energy Intelligence

Sir Peter Walters, former chairman, BP

Rawleigh Warner, former chairman, Mobil

Sir Denis Wright, former United Kingdom ambassador

Sheikh Ahmed Zaki Yamani, chairman, CGES (former oil minister, Saudi Arabia)

Mai Yamani, fellow, Royal Institute of International Affairs

Tammam Yamani

Daniel Yergin, chairman, CERA

Notes

Abbreviations are given below for private manuscript and public record collections that are frequently used in the notes. Also used extensively are abbreviations for the publications that Wanda Jablonski (WJ) was most closely associated with: *New York Journal of Commerce* (*JOC*), *Petroleum Week* (*PW*), and *Petroleum Intelligence Weekly* (*PIW*).

BGP   Brandon Grove Jr. Papers, Washington, DC
DYP   Daniel Yergin Papers, Cambridge, MA
JJM   John J. McCloy Papers, Amherst College Archives, Amherst, MA
MTP   Marshall Thomas Papers, Lavallette, NJ
NA   National Archives, Washington, DC
OMP   Onnic Marashian Papers, Oradell, NJ
PRO   Public Record Office, London, England
WEM   W. E. Mulligan Papers, Georgetown University Special Collections, Washington, DC
WJP   Wanda Jablonski Papers, Energy Intelligence, New York

PROLOGUE: WANDA'S LAST SCOOP

1. Thomas Wallin, interview by author, August 9, 1995; Wallin, telephone interview by author, February 23, 2002; Margaret Clarke, interview by author, October 8, 1996; *Petroleum Intelligence Weekly*, special supplement, April 6, 1992.
2. Daniel Yergin, *The Prize: The Epic Quest for Oil, Money, and Power* (New York: Simon & Schuster, 1991), 750–51, 760–61.
3. *New York Times*, October 30, 31, and November 3, 1986.
4. Wallin, interview, August 9, 1995.

5. Ibid.; Kenneth Miller, interviews by author, February 11, 1995, and December 2, 2001; Marshall Thomas, interview by author, June 15, 1995; Adam Smith, *Paper Money* (New York: Summit Books, 1981), 145; *New York Times*, October 30, 1986.

6. *PIW*, November 24, 1986; Sheikh Ahmed Zaki Yamani, interview by author, November 10, 1996.

7. Daniel Yergin, "Wanda as Matchmaker to OPEC's Founding Fathers," *PIW*, special supplement, April 6, 1992. Also see Yergin, *The Prize*, 517.

8. Anthony Sampson, *The Seven Sisters: The Great Oil Companies and the World They Shaped* (New York: Bantam Books, 1975), 7, 167; Yergin, *The Prize*, 503.

9. WJ to LeRoy Menzing, December 24, 1956, WJP; anonymous chief executive, interview by author, June 5, 1996.

10. James Tanner, interview by author, December 10, 1997; William Smith, interview by author, August 8, 1995; J. E. Hartshorn, interview by author, October 24, 1995; Lord Christopher Tugendhat, interview by author, November 14, 1996; Howard Macdonald, interview by author, November 12, 1996; Yamani, "Obituary—Wanda Jablonski," *Petroleum Review*, March 1992; *Times* (London), February 4, 1992.

1. PERIPATETIC YOUTH

1. Mary Jablonski, "Fifty Years of Our Life as I Saw It," undated (about 1960), unpublished essay, WJP; Dr. Vladimir Krcmery, letter to author, September 19, 1995; Krcmery, interview by author, July 12, 2003.

2. Victor S. Mamatey, "The Establishment of the Republic," *A History of the Czechoslovak Republic, 1918–1948*, ed. Victor S. Mamatey and Radomir Luza (Princeton, NJ: Princeton University Press, 1973), 6–8.

3. Krcmery, letter; Beverley Bowen Moeller, unpublished manuscript, 1976, WJP.

4. Krcmery, letter, May 18, 1998.

5. Ibid.; Krcmery, letter, September 19, 1998; John C. Jaqua, telephone interview by author, April 23, 1997; David Short, University of London, letter to author, February 21, 1998; Krcmery, letter, September 19, 1995; Josef Korbel, *Twentieth-Century Czechoslovakia: The Meanings of Its History* (New York: Columbia University Press, 1977), 103, 108.

6. Krcmery, letter, September 19, 1995; Mary Jablonski, "Fifty Years."

7. John C. Jaqua, letter to author, March 19, 1995; Moeller, manuscript; Mary

Jablonski, "Fifty Years"; Georgia Macris, interview by author, March 3, 1998; J. E. Hartshorn, interview by author, October 24, 1995.

8. Krcmery, letter, September 19, 1995; photos, WJP.

9. Mary Jablonski, "Fifty Years"; Jaqua, interview.

10. Mary Jablonski, "Fifty Years"; Krcmery, letter, September 19, 1995; Yergin, *The Prize*, 517.

11. "Eugene Jablonski Returns to Botany," *The Garden Journal: The New York Botanical Garden*, vol. 13, no. 3 (May–June 1963): 102–3; Mary Jablonski, "Fifty Years."

12. Ibid.

13. Mary Jablonski, "Fifty Years."

14. Ibid.; Krcmery, letter, September 19, 1995; Jaqua, interview.

15. Mary Jablonski, "Fifty Years."

16. Moeller, manuscript; "Jablonski Returns," *Garden Journal*; WJ, interview by Daniel Yergin, September 11, 1987, DYP.

17. Wanda Jablonski biography, McGraw-Hill Co. memo, undated; Arabian American Oil Co., *Arabian Sun and Flare*, February 1954; WJ to Charles Brooks, July 24, 1956, WJP.

18. Mary Jablonski, "Fifty Years"; "Jablonski Returns," *Garden Journal*; Moeller, manuscript; Mobil Corporation, "Socony Mobil Oil Company: A History in Brief," undated, WJP.

19. Mary Jablonski, "Fifty Years"; Yergin, *The Prize*, 195, 223.

20. Yergin, *The Prize*, 218–19; Mary Jablonski, "Fifty Years."

21. Moeller, manuscript; WJ to Brooks.

22. Moeller, manuscript; photos, WJP; Dr. Michael Osborne, interview by author, December 6, 1996; Joe Roeber, interview by author, October 30, 1995.

23. Mary Jablonski, "Fifty Years."

24. Ibid.; photos, WJP; Sarah Miller, interview by author, August 11, 1995; Moeller, manuscript; Eleanor Schwartz, interview by author, February 10, 1996.

25. Mary Jablonski, "Fifty Years"; photos, WJP; Moeller, manuscript.

26. Mary Jablonski, "Fifty Years"; St. Ann's Parochial School records, WJP; Moeller, manuscript; Jaqua, interview.

27. Mary Jablonski, "Fifty Years"; "Jablonski Returns," *Garden Journal*.

28. Mary Jablonski, "Fifty Years"; *Arabian Sun and Flare*, February 1954.

29. Mary Jablonski, "Fifty Years"; photos, WJP; Louis Moscatello, telephone interview by author, October 8, 1997.

30. Mary Jablonski, "Fifty Years"; Moeller, manuscript. Socony-Vacuum was

created in 1931 from the merger of Standard Oil of New York (Socony) and Vacuum Oil, later renamed Mobil Oil.

31. Yergin, *The Prize*, 203–5, 282; J. E. Hartshorn, *Oil Companies and Governments: An Account of the International Oil Industry and Its Political Environment* (London: Faber & Faber, 1962), 174–75.

32. Mary Jablonski, "Fifty Years"; Moeller, manuscript; photos, WJP; McGraw-Hill, *The Bulletin* (in-house publication), January 1957.

33. Sir Peter Walters, interview, November 14, 1996; Mary Jablonski, "Fifty Years"; Macris, interview; Ken Miller, interview by author, February 10, 1995; Sarah Miller, interview; Susan Habachy, telephone interview by author, October 31, 1996. Wanda also told some friends that she eloped as a teenager but that the marriage was annulled.

34. James Hunt, interview by author, June 5, 1996; Karol Krcmery, telephone interview by author, May 20, 1996. This cousin remembered hearing a version from Wanda when they were both quite young. Also see Yergin, *The Prize*, 517. Jablonski told Yergin that she traveled to Jerusalem by camel; her mother wrote that it was by car.

35. Mary Jablonski, "Fifty Years"; Jaqua, interview; Arthur G. Mills, telephone interview by author, November 16, 1996; Ambassador Brandon Grove Jr., interview by author, July 23, 1997; WJ, interview by Yergin; Krcmery, letter, September 19, 1995.

36. Sheikh Ahmed Zaki Yamani, interview by author, November 10, 1996.

37. Mary Jablonski, "Fifty Years."

38. Krcmery, letter, September 19, 1995; Mary Jablonski, "Fifty Years"; photos, WJP; Stefan Gregorovic to WJ, April 5, 1939, WJP (translation by author).

39. Mary Jablonski, "Fifty Years"; Moeller, manuscript; photos, WJP; WJ, interview by Yergin.

40. Grove Jr., interview.

41. Mary Jablonski, "Fifty Years"; Jaqua, letter, interview.

42. Jorg K. Hoensch, "The Slovak Republic, 1939–45," *History of the Czechoslovak Republic*, 272; Moeller, manuscript.

43. University of London, Matriculation Results; Cornell Admissions Office's Notice of Credit, WJP; Mary Jablonski, "Fifty Years"; Moeller, manuscript.

44. Moeller, manuscript. Wanda attended Harvard for one summer session.

45. Jaqua, interview; photos, WJP.

46. Jaqua, letter to author, April 25, 1995; Gregorovic, letter; photos, WJP.

47. Citizenship notice, undated, WJP; Moeller, manuscript; Osborne, interview; William Bland, telephone interview by author, March 2, 1998. Bland heard

this account from her in the mid-1950s. Federal Bureau of Investigation documents on Wanda, released in 1998 at the author's request, provide no additional information.

48. Denise Regan, interview by author, June 4, 1996; Jaqua, interview by author, April 29, 1997.

49. Schwartz, interview; Macris, interview by author, February 10, 1995, and telephone interviews by author, October 14, 1996, February 12, 1997, and March 3, 1998; Sister Mary Louise Sullivan, former president, Cabrini College, telephone interviews by author, October 24 and November 10, 1997; Hunt, interview; Krcmery, letter, September 19, 1995.

50. Columbia University records, WJP; Moeller, manuscript.

51. Ibid.; *Washington Post*, August 28, 1957; WJ, Deposition, August 2, 1988, U.S. District Court, Southern District of New York, 88 Civ. 3840, 119. She did not explain why she did not turn in her thesis.

2. "JUST CALL ME BILL"

1. Nan Robertson, *The Girls in the Balcony: Women, Men, and the New York Times* (New York: Fawcett Columbine, 1992), 109; Eileen Shanahan, interview by author, July 18, 1997.

2. *Frau im Leben* (Germany), August 1977; Beverley Bowen Moeller, manuscript, 1976, WJP; John Morton Blum, *V Was for Victory: Politics and American Culture during World War II* (New York: Harcourt Brace & Company, 1976), 95; Arabian American Oil Co., *Arabian Sun and Flare*, February 1954; *Washington Post*, August 28, 1957.

3. Chris Welles, "Fit or Unfit Business News," in *The Business Beat: Its Impact and Its Problems*, ed. William McPhatter (Indianapolis: Bobbs-Merrill, 1980), 21–23; Shanahan, interview; William Smith, interview by author, August 8, 1995; Kay Mills, *A Place in the News: From the Women's Pages to the Front Page* (New York: Columbia University Press, 1988), 255–56.

4. Edwin Emery and Michael Emery, *The Press and America: An Interpretive History of the Mass Media* (Englewood Cliffs, NJ: Prentice Hall, 1984), 488; James Tanner, interview, December 10, 1997; Aaron Riches, telephone interview by author, October 29, 1996.

5. Columbia University records; Moeller, manuscript; *Arabian Sun and Flare*, February 1954; McGraw-Hill, *The Bulletin*, January 1957: "Three weeks after being hired, she was writing stories. In 1945, after covering several other industries, she was moved to the oil desk—supposedly a temporary assignment, but she remained there as oil editor for ten years."

6. Robertson, *Girls*, 7–11, 100–103; Eleanor Schwartz, interview, February 10, 1996; Shanahan, interview (Shanahan was a leader in the women's discrimination suit against the *New York Times*); John C. Jaqua, interview, April 23, 1997; Eugene Bard to WJ, undated, 1977, WJP.

7. Schwartz, interview; Shanahan, interview.

8. Schwartz, interview; Jaqua, interview; Jaqua, letter to author, April 20, 1995.

9. Schwartz, interview; *Frau im Leben*, August 1977, 24; *Washington Post*, August 28, 1957.

10. *Washington Post*, August 28, 1957; *New York Journal of Commerce*, November 21 and 24, 1944; *New York Times*, August 19, 1979. Wanda implied that she got her oil-reporting job thanks to one particular scoop, but she conflated several stories into one.

11. *Oil & Gas Journal*, May 31, 1951, 120; Paul H. Frankel to WJ, January 3, 1948, WJP; M. A. Adelman, interview by author, July 10, 1996; William Bland, telephone interview by author, March 2, 1998; William Smith, interview; Tanner, interview; Lewis W. Foy, "How Business Assesses the Business Press," in *The Business Beat: Its Impact and Its Problems*, ed. William McPhatter (Indianapolis: Bobbs-Merrill, 1980), 40.

12. *JOC*, March 15, 1945.

13. Ibid.; S. Henle to WJ, March 16, 1945, WJP; Daniel Yergin, *The Prize: The Epic Quest for Oil, Money, and Power* (New York: Simon & Schuster, 1991), 395–403; Irvine H. Anderson, *Aramco, the United States, and Saudi Arabia: A Study of the Dynamics of Foreign Oil Policy, 1933–1950* (Princeton, NJ: Princeton University Press, 1981), 56–107, 124–25; Michael B. Stoff, *Oil, War, and American Security: The Search for a National Policy on Foreign Oil, 1941–1947* (New Haven, CT: Yale University Press, 1980), 151–77.

14. Dr. Michael Osborne, interview by author, December 6, 1996; Schwartz, interview; Mountain Lakes, NJ, newspaper clipping, wedding announcement, WJP.

15. Jaqua, interview; engagement (May 22, 1944) and wedding announcements, WJP.

16. Schwartz, interview; Jaqua, interview; Jaqua, letter; William Simpson to WJ, October 25, 1977, WJP.

17. Jaqua, interview; Jaqua, letter; Thelma Fissler to WJ, February 12, 1977, WJP.

18. Jaqua, letter; Schwartz, interview.

19. Ibid.; Jaqua, interviews, April 23 and 29, 1997.

20. Eugene Holman to WJ, September 5, 1946, WJP; Jaqua, interview, April 23, 1997; Schwartz, interview.

21. Schwartz, interview; anonymous chairman of a large independent oil

company, interview by author, June 5, 1996; Jaqua, letter; Riches, interview. (Riches attended a wedding party at Wanda's apartment.)

22. *JOC*, July 11, August 27, September 24, and October 19, 1945, February 18 and 27, 1946, and November 20, 1947.

23. Robertson, *Girls*, 63–70, 111; Kay Mills, *Place in the News*, 18–20, 50–57; Nancy Caldwell Sorel, *The Women Who Wrote the War* (New York: Harper-Collins, 2000), xiii–xiv, 27, 58–95, 113, 178–211, 242, 272, 294–316, 379–98.

24. Moeller, manuscript; Jaqua, interview, April 23, 1997; Jaqua, letter to author, March 19, 1995; Schwartz, interview; Shanahan, interview. Aside from Doran, other women reporters used their initials. When Shanahan joined in 1955, she broke that unwritten rule.

25. Shanahan, interview; Jaqua, interview, April 23, 1997. Numerous letters in Wanda's papers address her as "Mr. William Jablonski."

26. *Oil Light*, November 1950; WJ to James Stevens, editor, *Oil Light*, December 6, 1950, WJP.

27. Robertson, *Girls*, 64; Marilyn S. Greenwald, *A Woman of the Times: Journalism, Feminism, and the Career of Charlotte Curtis* (Athens: Ohio University Press, 1999), 62; Kay Mills, *Place in the News*, 65.

28. Robertson, *Girls*, 113; Shanahan, interview. Shanahan wanted to ask Kiplinger what kind of favors his male journalists "gave in return for the inside information *they* got," but she held her tongue.

29. Kay Mills, *Place in the News*, 60–61, 74; Robertson, *Girls*, 19, 64–67, 113; Shanahan, interview.

30. Shanahan, interview; Osborne, interview.

31. Robertson, *Girls*, 99–102; Kay Mills, *Place in the News*, 95–103.

32. Ron Chernow, *Titan: The Life of John D. Rockefeller, Sr.* (New York: Random House, 1998), 438; Yergin, *The Prize*, 105; Emery and Emery, *Press in America*, 322.

33. Edwin M. Yoder Jr., *Joe Alsop's Cold War: A Study of Journalistic Influence and Intrigue* (Chapel Hill: University of North Carolina Press, 1995), 19–20. Sylvia Porter wrote a syndicated column on financial matters. In the oil press, Ernestine Adams was an advertising copy and oil writer with Dallas-based *Petroleum Engineer*, and Ruth Sheldon Knowles was a freelance writer.

34. Richard H. K. Vietor, *Energy Policy in America since 1945: A Study of Business-Government Relations* (Cambridge: Cambridge University Press, 1984), 16–21, 35; Yergin, *The Prize*, 110, 226, 379.

35. Vietor, *Energy Policy*, 34–43; Yergin, *The Prize*, 226–27, 255–59; Stoff, *Oil, War*, 10–16; J. E. Hartshorn, *Oil Companies and Governments: An Account of*

the International Oil Industry and Its Political Environment (London: Faber & Faber, 1962), 135; M. A. Adelman, *The Genie out of the Bottle: World Oil since 1970* (Cambridge, MA: MIT Press, 1995), 42–50, 63.

36. Daniel Yergin and Joseph Stanislaw, *The Commanding Heights: The Battle Between Government and the Marketplace That Is Remaking the Modern World* (New York: Simon & Schuster, 1998), 49–50; Vietor, *Energy Policy*, 7; Stoff, *Oil, War*, 13.

37. Yergin, *The Prize*, 408–10; Adelman, *Genie*, 49.

38. *JOC*, November 20 and December 22, 1947.

39. Chernow, *Rockefeller*, 456–57; Yergin, *The Prize*, 102; "McCollum of Continental Oil," *Fortune*, September 1952, 126; "Rathbone of Jersey Standard," *Fortune*, May 1954, 118; Robert Sheehan, "Conoco's Widening World," *Fortune*, April 1964, 114–221; Bennett Wall, *Growth in a Changing Environment: A History of Standard Oil Company (New Jersey), Exxon Corporation, 1950–1975* (New York: McGraw-Hill, 1988), xxiii–xiv; *Business Week*, July 2, 1960, 42; *JOC*, September 17 and October 15, 1948.

40. *JOC*, February 6, and December 15 and 22, 1948.

41. Jaqua, interview, April 29, 1997.

42. *JOC*, September 17, 1948.

43. Sidney Swensrud to WJ, November 3, 1948, WJP; Bruce Brown to WJ, November 8, 1948, WJP; Monroe Rathbone to WJ, December 8, 1948, WJP; James Crayhon to WJ, January 7, 1949, WJP; *International Oil Worker*, September 27, 1948. The *Oil Worker* criticized "Mr." Jablonksi for being an industry apologist.

44. *JOC*, October 29, 1948; Brewster Jennings to WJ, November 4, 1948, WJP; Brown Meece to WJ, November 4, 1948, WJP; Rathbone to WJ, December 8, 1948, WJP; Leonard McCollum to WJ, November 4, 1948, WJP; Harry J. Kennedy to McCollum, December 6, 1948, WJP; Emery and Emery, *Press and America*, 521.

45. *JOC*, October 22, 1948; James Terry Duce to WJ, November 1, 1948, WJP; Swensrud to WJ, November 3, 1948, WJP; McCollum to WJ, WJP; Everette DeGolyer to WJ, November 24, 1948, WJP.

46. Jaqua, interview, April 23, 1997; Dame Edith Penrose, interview by author, October 27, 1995; Burton I. Kaufman, *The Oil Cartel Case: A Documentary Study of Antitrust Activity in the Cold War Era* (Westport, CT: Greenwood Press, 1978), 162–70.

47. *JOC*, November 20 to December 22, 1947; Wallin, interview, August 8, 1995; William Lindenmuth, interview by author, December 7, 1996.

48. *JOC*, December 12, 1946.

49. Ibid.; Wallin, interview, August 9, 1995.

50. *JOC*, May 26 and 28, 1947.

51. Jaqua, interview, April 23 and 29, 1997; Schwartz, interview.

52. Yergin, *The Prize*, 271–77, 436; Stephen G. Rabe, *The Road to OPEC: United States Relations with Venezuela, 1919–1976* (Austin: University of Texas Press, 1982), 104; *Fortune*, February 1949, 178–79.

53. Yergin, *The Prize*, 435–37; Rabe, *Road to OPEC*, 94–102.

54. *JOC*, January 10, 1946.

55. *JOC*, May 2 and December 31, 1946, July 29, 1947, and November 23 and 26, 1948. She quoted one American producer, William F. Buckley, who predicted the Aramco deal would be disastrous for Venezuelan oil.

56. John Pearson, interview by author, June 6, 1996; Shanahan, interview; Ruth Pearson, telephone interview by author, November 7, 1996; Ruth Pearson, interview by author, December 3, 2001.

57. J. L. Waldman (editor, *Caracas Journal*) to WJ, April 15, 1948, WJP; *JOC*, February 20 to March 4, 1948.

58. *JOC*, February 20, 1948; Keith Hutchinson, "Everybody's Business: *Sembra el Petroleo*," *Nation*, March 13, 1948, 305; Waldman to WJ. The English-language *Caracas Journal* reprinted her column.

59. *JOC*, March 4, 1948; Alirio Parra, former oil minister of Venezuela, interview by author, November 1, 1995.

60. *JOC*, March 4, 1948.

61. Perez Alfonso to WJ, March 22, 1948, WJP.

62. Robert Siegel to WJ, March 15, 1948, WJP.

63. Frank W. Abrams to WJ, March 22, 1948, WJP; John Pearson, interview; Tanner, interview; Wall, *Growth*, xxiii.

64. John Pearson, interview; Eugene Holman to WJ, September 5, 1946, WJP; *JOC*, March 15, 1945; Wall, *Growth*, xxii, xxiv, 11–13.

65. K. E. ("Ted") Cook to WJ, April 16, 1948, WJP; Abrams to WJ.

66. Edward D. Mellinger to Annette Chaikin, September 29, 1950, WJP.

67. *JOC*, March 29, 1948.

68. George F. Meredith, executive director of the U.S. Senate Special Committee to Study Problems of American Small Business, to WJ, April 19, 1948, WJP; WJ to Meredith, April 23, 1948, WJP; *JOC*, June 18, 1948, and September 11, October 9, November 10, and December 4 and 15, 1952.

69. Bryant Putney to WJ, June 12, 1952, WJP; A. P. Frame to WJ, April 27, 1951, WJP; Deputy PAD Administrator Bruce Brown to WJ, April 25, 1951, WJP;

U.S. Congress, House of Representatives, Committee on Interstate and Foreign Commerce, Petroleum Study (Gasoline and Oil Price Increases): Hearings, July 1–23, 1953, 167; Martin Linsky, *Impact: How the Press Affects Federal Policymaking* (New York: W.W. Norton, 1986), 102–11.

70. House Commerce Committee, Petroleum Study Hearings, 2, 649.

71. *JOC*, June 9, 1953.

72. Macris interview, February 12, 1995; John Lichtblau, interview by author, February 11, 1995; *JOC*, July 7, 1953.

73. House Commerce Committee, Petroleum Study Hearings, 324–27; Serge Jurenev to WJ, July 15 and August 17, 1953, WJP.

74. *JOC*, July 14 and 23, 1953.

75. *JOC*, July 23, 1953.

76. John Boatwright, chief economist, Standard Oil of Indiana, to WJ, August 20, 1953, WJP; House Commerce Committee, *Petroleum Study: Hearings*, 648, and 58, 167, 174, 649, 744; John Lichtblau, untitled note, *PIW*, special supplement, April 6, 1992. Lichtblau recalled that an irate oil producer dismissed an article by Wanda at a Washington hearing because it was written by "a girl."

77. *Ethyl News*, May–June 1953; Margaret Clarke, interview by author, October 8, 1996; Jaqua, interview, April 23, 1997.

3. IRANIAN INTRIGUE

1. *JOC*, February 9, 1954; Denis Wright, interview by author, October 31, 1995, WJP; Wright to WJ, April 21, 1954, WJP; Parviz Mina, interview by author, November 15, 1996.

2. Georgia Macris, interview by author, October 13, 1996; *JOC*, December 22, 1953.

3. *Collier's*, January 21, 1955; WJ appointment diaries, 1953–54, WJP; Manucher Farmanfarmaian, interview by author, December 4, 1996; Rostam ("Mike") Bayandor, interview by author, November 25, 1995.

4. Wright, interview; *JOC*, January 6 and February 8, 11, 16, and 25, 1954.

5. Daniel Yergin, *The Prize: The Epic Quest for Oil, Money, and Power* (New York: Simon & Schuster, 1991), 451–58; Stephen Kinzer, *All the Shah's Men: An American Coup and the Roots of Middle East Terror* (Hoboken, NJ: John Wiley & Sons, 2003), 2, 76–82, 91.

6. Mostafa Elm, *Oil, Power, and Principle: Iran's Oil Nationalization and Its*

Aftermath (Syracuse, NY: Syracuse University Press, 1992), 144–48; Kinzer, *All the Shah's Men*, 167–92; Yergin, *The Prize*, 468–70. Production fell to 22,000 barrels a day from 650,000 in 1950.

7. Yergin, *The Prize*, 470–76.

8. Harlan Cleveland, "Oil, Blood and Politics: Our Next Move in Iran," *Reporter*, November 10, 1953, 11–19; John Lewis Gaddis, *Strategies of Containment: A Critical Appraisal of Postwar American National Security Policy* (New York: Oxford University Press, 1982), 30.

9. Farmanfarmaian, interview; *JOC*, February 9, 1954.

10. Mina, interview; WJ appointment diaries, 1953–54, WJP.

11. Mina, interview; J. H. Bamberg, *The Anglo-Iranian Years, 1928–1954*, Vol. 2, *The History of the British Petroleum Company* (Cambridge: Cambridge University Press, 1994), 90–95, 105.

12. *Wall Street Journal*, April 7, 1951; Elm, *Oil, Power*, 88–89, 102–3, 159, 227; Kinzer, *All the Shah's Men*, 95–96; Sir Oliver Franks, Foreign Office, "Survey of American Press," *British Foreign and State Papers*, 1953, F0371/91615, PRO.

13. *Wall Street Journal*, May 23 and June 5, 9, 11, 21, and 23, 1951.

14. *Wall Street Journal*, June 5, 1951.

15 *Wall Street Journal*, July 3 and 6, 1951, and October 1 and 14, 1953; Elm, *Oil, Power*, 103; Philip Noel-Baker to Ernest Bevin, Foreign Office 371/91628, November 15, 1950, PRO.

16. *Business Week*, June 30, August 4, and September 1, 1951, and August 29 and October 24, 1953; *Fortune*, October 1954, 208, and May 1955, 85; *Oil & Gas Journal*, October 18, 1951, 69–71.

17. *JOC*, May 3 and 17, June 14, 21, and 28, July 18 and 19, August 16, and November 1 and 8, 1951; and *JOC*, October 8, 1953.

18. *JOC*, December 28 and 30, 1953.

19. J. T. Fearnley, British embassy, Tehran, to Foreign Office, January 9, 1954, PD 230/99/4, PRO; Wright, "The Memoirs of Sir Denis Wright, 1911–1971," unpublished, Sir Denis Wright Papers, Haddenham, Bucks (UK), Vol. 1, 221; Wright, letter to author, May 11, 1995; Walter Levy, interview by author, August 9, 1995; P. T. Hart to B. R. Byroade, July 10, 1953, 59 880.2553/7–1053, NA; Loy Henderson to John Jernegan, November 12, 1953, 880.2553/11–1253, Record Group 59, Department of State, National Archives.

20. Loy Henderson, cables to State Department: December 16, 1953, 888.2553/12–1653, December 23, 1953, 888.2553/12353, December 29, 1953, RG 59, NA; *News Report*, United States Information Agency, Iran, December 31, 1953, citing Wanda's December 30, 1953, *JOC* story; Arabian American Oil Co., *Arabian Sun and Flare*, February 1954.

21. Wright, interview; WJ's notes, translations of Iranian news articles, photos, WJP; Macris, interview, February 10, 1995.

22. *Ettelaat* (Tehran), January 19, 1954; *Dad* (Tehran), January 19, 1954.

23. *Seday-e-Vatan* (Tehran), January 20, 1954; *Farman* (Tehran), January 17, 1954 (based on exclusive interview with WJ); *Ettelaat*, January 19, 1954; *Mosavar* (Tehran), January 22, 1954; unidentified translations and WJ annotations on newspaper clips, WJP; S. K. Kazerooni to WJ, April 23, 1954, WJP.

24. *JOC*, February 11, 1954.

25. Wright to WJ, April 21, 1954, WJP; Wright, interview; Wright, "Memoirs," 221.

26. Ibid.; James Bill, *The Eagle and the Lion: The Tragedy of American-Iranian Relations* (New Haven, CT: Yale University Press, 1988), 106.

27. James Bill, *Eagle and the Lion*, 107; Wright, interview; Naficy to WJ, April 5, 1954, WJP; Fathollah Naficy, telephone interview by author, June 1, 1996; Farrokh Najmabadi, interview by author, June 14, 1996.

28. Fearnley, British embassy, Tehran, to Foreign Office, January 9, 1954, PD 230/99/4, PRO; Farmanfarmaian, interview.

29. Hussein Ala to WJ, February 4, 1954, WJP.

30. Najmabadi, interview; Farmanfarmaian, interview.

31. Ibid.; Mina, interview. Also see essays about Middle Eastern women in Mai Yamani, ed., *Feminism and Islam: Legal and Literary Perspectives* (Reading, UK: Ithaca Press, 1996).

32. Manucher Farmanfarmaian and Roxane Farmanfarmaian, *Blood and Oil: Memoirs of a Persian Prince* (New York: Random House, 1997), 307; photos, WJP; Brandon Grove Jr., interview by author, July 23, 1997.

33. *New York Times*, January 6, 1954; *JOC*, January 19, 1954.

34. Henderson to State Department, 888.2553/12–2953, NA; Edwin Moline to State Department, 880.2553/12–3053, RG 59, NA; John H. Brook to Angus Beckett, Ministry of Fuel and Power, January 6, 1954, and December 30, 1953, PD 230/99/4, PRO. Hoover did not leave his own record of their contacts. His papers were destroyed upon his death in 1969, according to Herbert ("Pete") Hoover III, e-mail to author, June 7, 1997.

35. Wright, cable to Foreign Office, January 7, 1954, FO37/100059, PRO. All Wright could glean about Abadan in January 1954 was from secondhand accounts of a Dane and a Swede.

36. *JOC*, February 19, 1954; Flora Lewis, *New York Times Magazine*, July 8, 1951, 9; Kay Mills, *A Place in the News: From the Women's Pages to the Front Page* (New York: Columbia University Press, 1988), 200, 206.

37. *JOC*, February 4, 16, and 23, 1954.

38. *JOC*, February 23, 25, and 26, 1954.

39. *JOC*, February 26, 1954.

40. *JOC*, February 25, 1954; Farmanfarmaian and Farmanfarmaian, *Blood and Oil*, 184–85.

41. *JOC*, February 19, 1954.

42. Kazerooni to WJ, WJP; A. Gosem Kheradjan to WJ, March 28, 1954; M. A. Hirmand to WJ, March 24, 1954, WJP.

43. Anna Rubino, "Recollections of Life at PIW through the Years," *PIW*, special supplement, April 6, 1992; Ken Miller, interview by author, February 10, 1995; Macris, interview, February 10, 1995; J. E. Hartshorn, interview by author, October 24, 1995; Margaret Clarke, interview by author, October 8, 1996; Wright, interview. Wright recalled that WJ said "the Iranians grew suspicious of her activities and asked her to leave the country."

44. Jawad Khalili to WJ, March 25, 1954, WJP; L. T. Jordan, cable to WJ, February 8, 1954, WJP; C. W. Hamilton, Gulf Oil vice president, to WJ, April 8, 1954, WJP.

45. *JOC*, March 11, 1954; Sami Nasr to WJ, April 16, 1954, WJP.

46. James MacPherson to WJ, April 26 and July 12, 1954, WJP.

47. *JOC*, March 4, 1954; *Arabian Sun and Flare*, February 1954; WJ story cited in *Petroleum Times*, March 19, 1954.

48. James Critchfield, Central Intelligence Agency's national energy intelligence officer, interview by author, August 5, 1995; Lois Critchfield, letter to author, March 9, 1998. In response to a Freedom of Information Act request on February 3, 1995, the CIA stated that it could neither confirm nor deny any confidential relationship with WJ and that no records could be located. CIA, letter to author, April 14, 1998. Peter Grose, *Gentleman Spy: The Life of Allen Dulles* (Boston: Houghton Mifflin, 1994), 387; Ronald Steel, *Walter Lippmann and the American Century* (New York: Little, Brown, 1980), 510; Edwin Yoder Jr., *Joe Alsop's Cold War: A Study of Journalistic Influence and Intrigue* (Chapel Hill: University of North Carolina Press, 1995), 151. This practice ended abruptly in the early 1970s because of Vietnam-era concerns about journalistic ethics.

49. John C. Jaqua, interview, April 29, 1997; James Critchfield, interview. The CIA's New York office did not handle covert agents.

50. *Arabian Sun and Flare*, February 1954; Helen Brown to WJ, March 11, 1954, WJP; Jaqua, interview, April 29, 1997.

51. Jaqua, interview, April 23, 1997; Nan Robertson, *The Girls in the Balcony: Women, Men, and the New York Times* (New York: Fawcett Columbine, 1992), 65–70; C. B. Squire, telephone interview by author, August 29, 1997; William Smith, interview.

52. Jaqua, interview, April 23, 1997; Jaqua to author, April 25, 1995; Schwartz, interview; Macris, interview, July 20, 1994; John Buckley, interview by author, July 7, 1996; James Hunt, interview by author, June 2, 1996.

53. Anonymous chairman of a large independent oil company, interview; Levy, interview; Farmanfarmaian, interview; Sir Eric Drake, interview by author, November 2, 1995.

54. Jaqua, interview, April 23, 1997; Macris, interview, July 20, 1994; Clarke, interview; Schwartz, interview.

55. Squire, letter to author, August 7, 1997.

56. Elm, *Oil, Power*, 330–31; *Business Week*, August 7, 1954.

57. *JOC*, August 5, 1954; *Collier's*, January 21, 1955.

58. *Collier's*, January 21, 1955

59. Ibid.; James Bill, *Eagle and the Lion*, 107.

60. *Collier's*, January 21, 1955.

61. WJ to Hoover, January 6, 1955, WJP; Hoover to WJ, January 7, 1955, WJP; WJ to Ala, January 6, 1955, WJP; WJ to Loy Henderson, January 6, 1955, WJP; WJ to T. Reiber, January 6, 1955, WJP; WJ to Denis Wright, January 6, 1955, WJP; Allen W. Dulles to WJ, January 13, 1955, WJP.

62. *Collier's*, January 21, 1955.

4. SAVORING THE BEDOUIN BREW

1. WJ to LeRoy Menzing, December 6 and 7, 1956, WJP.

2. Diane Kunz, *The Economic Diplomacy of the Suez Crisis* (Chapel Hill: University of North Carolina Press, 1991); Daniel Yergin, *The Prize: The Epic Quest for Oil, Money, and Power* (New York: Simon & Schuster, 1991), 479–500.

3. WJ to Menzing, December 7, 1956.

4. *PW*, December 14, 1956.

5. *PW*, February 22, 1957.

6. McGraw-Hill, press release, undated, 1957; Sami Nasr to WJ, May 8, 1961, WJP.

7. M.A. Adelman, interview by author, July 10, 1996; Adelman to Edward Morse, April 12, 1992, WJP.

8. William Bland, telephone interview by author, March 2, 1998.

9. Adelman, interview; Georgia Macris, interview by author, February 10, 1995; Macris, memo to Onnic Marashian, March 29, 1957, OMP; C.B. Squire, telephone interview by author, August 29, 1997; Anna Rubino, "Recollections

of Life at PIW through the Years," *PIW*, special supplement, April 6, 1992.

10. Squire, interview; Bland, interview.

11. Macris, interview; Harriet Costykian, interview by author, February 12, 1996; WJ to James MacPherson, October 18, 1957, WJP.

12. Yergin, *The Prize*, 412, 435–37; Arabian American Oil Company, *Aramco Handbook: Oil and the Middle East* (Arabian American Oil Company: Dhahran, Saudi Arabia, 1968), 135; Irvine H. Anderson, *Aramco, the United States, and Saudi Arabia: A Study of the Dynamics of Foreign Oil Policy, 1933–1950* (Princeton, NJ: Princeton University Press, 1981), 108–15, 179–97, 185; Wallace Stegner, *Discovery: The Search for Arabian Oil* (Beirut: Middle East Export Press, 1971); Richard Nolte, "A Tale of Two Cities," *Fieldstaff Reports: Asia*, Vol. 20, no. 1 (Hanover, NH: American Universities Field Staff, 1977), 5; Baldo Marinovic interview in Carole Hicke, ed., *American Perspectives of Aramco, the Saudi-Arabian Oil-Producing Company, 1930s to 1980s*, an oral history conducted 1992–1993, Regional Oral History Office, Bancroft Library, University of California–Berkeley, 1995, 272–77; Said K. Aburish, *The Rise, Corruption, and Coming Fall of the House of Saud* (New York: St. Martin's Press, 1994), 38–41, 277.

13. Yergin, *The Prize*, 447; Anderson, *Aramco*, 183, 202–3; Michael B. Stoff, *Oil, War, and American Security: The Search for a National Policy on Foreign Oil, 1941–1947* (New Haven, CT: Yale University Press, 1980), 215.

14. Macris, interview, February 12, 1995; Arabian American Oil Co., *Arabian Sun and Flare*, June 22, 1949; Anderson, *Aramco*, 110; Francis E. Meloy Jr. to Secretary of State, December 12, 1948, 890F.6363/12–1248, RG 59, NA; Parker T. Hart to Secretary of State, July 2, 1949, U.S. Congress, Senate, Committee on Foreign Relations, Subcommittee on Multinational Corporations, *Multinational Corporations and United States Foreign Policy, Hearings*, part 7, 85–89; MacPherson to U.S. Ambassador J. Rives Childs, June 11, 1949, U.S. Congress, *Multinational Corporations*, part 7, 94–95; Robert Vitalis, *America's Kingdom: Mythmaking on the Saudi Oil Frontier* (Stanford, CA: Stanford University Press, 2007), 106.

15. MacPherson to WJ, February 22 and June 6, 1955, WJP.

16. *PW*, July 8, 1955.

17. MacPherson to WJ, July 2 and 17, 1955, WJP.

18. MacPherson to WJ, November 27, 1955, WJP; Grace MacPherson to James MacPherson, November 23, 1955, WJP; WJ to MacPherson, December 9, 1955, WJP.

19. *PW*, August 26, 1955; Sir Francis Hopwood to WJ, September 21, 1955, WJP. Hopwood did not like her column but said it was "meticulously accurate."

20. WJ, interview by Yergin, September 11, 1987, DYP; WJ to Menzing, December 11, 1956, WJP.

21. *PW*, May 25 and August 10, 1956; S. A. Swensrud to John Foster Dulles, August 10, 1956, WJP.

22. WJ to Menzing, December 12, 1956, WJP.

23. *PW*, January 11, 1957.

24. WJ to Menzing, December 12, 1956; Squire, interview; Squire, letter to author, August 7, 1997.

25. WJ to Menzing, December 12, 1956; WJ, interview by Yergin.

26. WJ to Menzing, December 12, 1956; Miles Copeland, *The Game of Nations: The Amorality of Power Politics* (New York: Simon & Schuster, 1969), 70, 166; Onnic Marashian, interviews by author, February 9 and June 4, 1996.

27. Marashian, interviews; Marashian to WJ, October 25, 1956, OMP.

28. Marashian, interviews; WJ to Menzing, December 12, 1956.

29. Ibid.

30. WJ to Menzing, December 12, 1956; *PW*, November 9, 1956; McGraw-Hill, *The Bulletin*, January 1957, 21.

31. *PW*, December 13, 1956.

32. Phebe Marr, telephone interview by author, May 5, 1999; WJ to Menzing, December 14, 1956, WJP.

33. WJ to Menzing, December 24, 1956, WJP; WJ, interview by Yergin.

34. Ibid.; Nolte, "Tale," 16; Richard Nolte, interview by author, February 8, 1996; Marr, interview.

35. WJ to Menzing, December 24, 1956, WJP.

36. Ibid.; Marr, interview; Ellen Spears interview, *American Perspectives*, 559–60.

37. Marr, interview; Ambassador Parker Hart, *Oral History Interviews*, January 27, 1989, Georgetown University Special Collections.

38. Jack Butler, interview by author, June 4, 1996.

39. *Mirror-News* (Los Angeles), June 14, 1957; WJ, interview by Yergin.

40. *Mirror-News*, June 14, 1957.

41. *Daily Journal* (Caracas), December 6, 1957.

42. *Mirror-News*, June 14, 1957.

43. Ibid.; Yergin, *The Prize*, 517; W. E. Turner to WJ, April 16, 1957, WJP; Knox Bourne to Harry Waddell, June 17, 1957, WJP; Leslie Recordings to WJ, June 17, 1957, WJP; *Daily Mirror* (Caracas), December 6, 1957.

44. *PW*, January 25 and February 15, 1957.

45. Peter Kurth, *American Cassandra: The Life of Dorothy Thompson* (Boston: Little, Brown, 1990), 382–85, 428.

46. *PW*, January 18, 1957.

47. Ibid.; *New York Post,* January 21, 1957; James Prescott to WJ, January 23, 1957, WJP.

48. *PW,* January 25 and February 8, 1957.

49. *Time,* January 28, 1957; *PW,* February 8, 1957; David Holden and Richard Johns, *The House of Saud: The Rise and Rule of the Most Powerful Dynasty in the Arab World* (New York: Holt, Rinehart & Winston, 1981), 178–99; Vitalis, *America's Kingdom,* 200–10.

50. Rubino, "Recollections."

51. *PW,* February 15, 1957.

52. WJ to Menzing, December 24, 1956, WJP; *PW,* February 15, 1957; Marshall Thomas, "Highlights from Marshall Thomas," *PIW,* special supplement, April 6, 1992; *Daily Journal,* December 6, 1957.

53. *PW,* April 12, 1957.

54. *PW,* March 8, 1957; Ruth Pearson, *Daily Journal,* December 6, 1957.

55. Robert D. Kaplan, *The Arabists: The Romance of an American Elite* (New York: Free Press, 1995), 59–60; Freya Stark, *The Freya Stark Story* (New York: Coward-McCann, 1953), 301–2.

56. Geoffrey Herridge to Brandon Grove, August 21, 1956, WJP.

57. WJ to C. A. P. Southwell, October 11, 1956, WJP; William Fraser to WJ, October 29, 1956, WJP; WJ to Fraser, November 26, 1956, WJP; WJ to Wright, October 30, 1956, WJP; British Petroleum memo in Fraser letter to Geoffrey Stockwell, November 20, 1956, WJP; WJ to L. Astley-Bell, November 26, 1956, WJP.

58. McGraw-Hill, press release, undated, 1957, WJP; Paul Ensor, telephone interview by author, November 17, 1996; Clarke, interview.

59. Ensor, interview.

60. Ibid.; Ensor, telephone interview by author, October 20, 1996.

61. Clarke, interview.

62. *PW,* April 12, 1957.

63. Ensor, phone interview, November 17, 1996; WJ notes, WJ appointment diary, 1957, WJP.

64. *PW,* March 15, 1957.

65. Ibid.; Ensor, interview, October 20, 1996.

66. PBS, "The Oil Kingdoms," Vol. 1, Jo Franklin-Trout, producer, 1985.

67. Ensor, interview, November 17, 1996.

68. *PW,* March 15, 1957.

69. Frank R. Wyant, *The United States, OPEC, and Multinational Oil* (Lexington, MA: Lexington Books, 1977), 65; *PW,* March 22, April 12 and 26, and May 3, 1957.

70. *PW*, March 22 and May 17, 1957.

71. *PW*, March 15, 22, April 12, 26, and May 3, 17, 1957.

72. T. A. Hillyard to WJ, June 5, 1957, WJP; Susan Hillyard to WJ, September 27, 1957, WJP; Ensor, interview.

73. Holden and Johns, *House of Saud*, 193–94. See minimal coverage in January and February 1957 in *New York Times, Wall Street Journal, Fortune*, and *Business Week*.

74. Ibid.

75. *PW*, March 15, 1957.

76. WJ appointment diary, 1957, WJP; *PW*, March 15 and April 5, 1957.

77. *PW*, March 15, 1957.

78. Ibid.

79. *PW*, April 19, 1957.

80. WJ, telegram to Gamal Abdel Nasser, March 1957, OMP; William Beard Jr. to WJ, March 8, 1957, WJP. Associated Business Publications was later renamed American Business Media.

81. Macris, interview, February 10, 1995; Dale Carnegie, certificate, May 15, 1956, WJP.

82. Schwartz, interview; James Critchfield, interview, August 5, 1995; Michael Ameen, interview by author, November 25, 1997; Butler, interview.

83. John C. Jaqua, telephone interview by author, April 23, 1997; Schwartz, interview; Macris, interview by author, July 20, 1994.

84. W. E. Turner to WJ, April 16, 1957, WJP; Convention Chairman Gilbert Palmer to WJ, May 27, 1957, WJP; J. E. MacArthur, memo to LeRoy Menzing, May 27, 1957, WJP; Knox Bourne, memo to Harry Waddell, June 17, 1957, WJP; Macris, interview, February 10, 1994.

85. Standard Oil of California, memo, "Notes on the talk by Miss Wanda Jablonski at the World Affairs Council luncheon," June 11, 1957, WJP.

5. THE ADVENT OF OIL NATIONALISM

1. Sheikh Ahmed Zaki Yamani, interview by author, November 10, 1996; Mike Ameen, interview by author, November 25, 1997.

2. WJ to Menzing, December 24, 1956, WJP; *PW*, February 22, 1957; Ameen, interview; Yamani, interview; Jack Butler, interview by author, June 4, 1996.

3. *PW*, February 22, 1957; Butler, interview; Ameen, interview; Abdullah Tariki to WJ, September 19, 1957, WJP; WJ to Edward Brown, November 25, 1957, WJP.

4. *PW*, October 18, 1957.

5. Tariki to WJ, WJP; WJ to Tariki, October 14, 1957, WJP.

6. WJ to James MacPherson, October 18, 1957, WJP; *PW*, October 25, 1957.

7. Brown to WJ, December 2, 1957, WJP; WJ to Robert Brougham (Aramco
vice president), April 14, 1958, WJP; David Holden and Richard Johns,
*The House of Saud: The Rise and Rule of the Most Powerful Dynasty in the
Arab World* (New York: Holt, Rinehart & Winston, 1981), 199; *New York
Times*, April 3 and June 4, 1958; *Wall Street Journal*, January 19, 1957, and
March 26, 1958.

8. *PW*, June 6, 1958.

9. *PW*, June 20, 1958.

10. Ibid.

11. WJ, interview by Yergin, September 11, 1987, DYP.

12. Feisal Mazidi, interview by author, October 31, 1995; Ambassador Richard
Nolte, interview by author, February 8, 1996.

13. *Wall Street Journal*, August 13 and 19, 1958; *Time*, October 27, 1958; *New York
Times*, March 7 and October 31, 1957, June 4, 1958 (AP report). Tariki was not
mentioned in *Business Week* or *Newsweek* until 1960.

14. *Fortune*, May and November 1958. *Fortune* failed to note the irony: U.S.
production controls were acceptable, but foreign restraints were not.

15. *PW*, February 22, 1957; U.S. embassy, Jeddah, memo to Department of State,
June 5, 1957, 880.2553/6–557, RG 59, NA; Ameen, interview by author,
September 20, 1995, and telephone interview by author, November 25,
1997.

16. Ameen, interviews.

17. J. E. Hartshorn, interview by author, October 24, 1995; Nolte, interview;
M. A. Adelman, interview by author, July 10, 1996; Richard Nolte, "A Tale of
Two Cities," *Fieldstaff Reports: Asia*, Vol. 20, no. 1 (Hanover, NH: American
Universities Field Staff, 1977); WJ to E. W. Tatge, September 16, 1957, WJP.

18. Ameen, interviews; Central Intelligence Agency, "Saudi Arabia: Abdullah
Ibn Hamud al-Tariqi," February 26, 1970, *Declassified Documents Reference
System* (Washington, DC: Carrollton, 1977–81); Peter Speers, interview in
Carole Hicke, ed., *American Perspectives of Aramco, the Saudi-Arabian Oil-
Producing Company, 1930s to 1980s*, Regional Oral History Office, Bancroft
Library, University of California–Berkeley, 1995, 491.

19. CIA, "Saudi Arabia: Abdullah Ibn Hamud al-Tariqi"; Anthony Cave Brown,
Oil, God, and Gold: The Story of Aramco and the Saudi Kings (Boston:
Houghton Mifflin, 1999), 152; Ameen, interviews.

20. John C. Jaqua, interview by author, April 23, 1997; Michael Saltser (WJ's

lawyer and executor), interview by author, May 30, 1995; Eleanor Schwartz, interview by author, February 10, 1996; Margaret Clarke, interview by author, October 8, 1996; Harriet Costikyan, interview, June 4, 1996.

21. Onnic Marashian, interview by author, June 4, 1996.

22. WJ to Marashian, July 24, 1956, OMP; Macris to Marashian, January 21, 1958 and October 16, 1956, OMP; Marashian to Macris, October 10, 1956, OMP; Macris to Marashian, October 13, 1959, OMP; WJ to Marashian, October 21, 1959, OMP; Marashian to WJ, October 29, 1959, OMP.

23. Butler, interview; Ameen, interview, November 25, 1997; John W. Pendleton, Special Committee, "Confidential Memorandum to the Files," December 30, 1957, Box 2, File 57, WEM; K. R. Webster to G. E. Mandis, March 6, 1957, Box 2, File 55, WEM.

24. Yamani, interview.

25. Ibid.; Butler, interview.

26. A. Franklyn Williams (British embassy, Washington) to Keith L. Stock (Ministry of Power, London), June 23, 1958, POWE 33/2200, PRO.

27. Ibid.; "Rathbone of Jersey Standard," *Fortune*, May 1954; Daniel Yergin, *The Prize: The Epic Quest for Oil, Money, and Power* (New York: Simon & Schuster, 1991), 521; WJ, interview by Yergin.

28. Williams to Stock, PRO; Albert Nickerson to WJ, June 25, 1958, WJP; George Parkhurst to WJ, June 4, 1958, WJP; William Whiteford to WJ, June 9, 1958, WJP.

29. Rathbone to WJ, August 28, 1958, WJP; Howard Page to WJ, September 3, 1958, WJP; John Loudon to WJ, September 19, 1958, WJP; Nickerson to WJ, September 17, 1958, WJP; Marashian, interview, June 4, 1996; Pierre Terzian, *OPEC: The Inside Story*, trans. Michael Pallis (London: Zed Books, 1985), 85. Page "couldn't stand" Tariki, Terzian wrote.

30. *PW*, September 20, November 8, and December 13 and 20, 1957; *PW*, September 19, 1958; Macris to Marashian, October 16, 1956, OMP; Marashian to Macris, December 13, 1957, OMP; Macris to Marashian, December 19, 1957, OMP; Marashian to Macris, June 4, 1958, OMP; Brown to WJ, WJP.

31. Macris to Marashian, May 28, 1958, OMP.

32. W. K. Beard to WJ, March 12, 1958; Marschal Rothe Jr. to WJ, February 5, 1958, WJP; H. B. Bullard to WJ, October 15, 1957, WJP; Marian Young (Martha Deane) to WJ, February 14, 1958, WJP; Harry Lewis to WJ, October 18, 1957, WJP; *New York Times*, October 18, 1957; *PW*, March 21, 1958.

33. Stephen G. Rabe, *The Road to OPEC: United States Relations with Venezuela, 1919–1976* (Austin: University of Texas Press, 1982), 117–32; *PW*,

January 17 and April 4, 1958; Franklin Tugwell, *The Politics of Oil in Venezuela* (Stanford, CA: Stanford University Press, 1975), 183.

34. Ruth Pearson, telephone interview by author, November 7, 1996; *Daily Journal* (Caracas), December 6, 1957.

35. Ruth Pearson, interview; Dame Edith Penrose, interview by author, October 27, 1995; *Panorama* (Caracas), December 8, 1957; *Diario de Presidente* (Maracaibo), December 1957, WJP.

36. John Pearson, interview by author, June 6, 1996.

37. Rabe, *Road to OPEC*, 133–34; PW, January 17, 1958.

38. U.S. embassy, Caracas, memo to Department of State, January 16, 1958, 831.2553/1–1658, RG 59, NA; *PW*, January 24 and February 28, 1958.

39. *PW*, January 31, 1958.

40. *PW*, March 7, 1958.

41. *PW*, March 21, 1958.

42. *PW*, March 14, 1958.

43. *PW*, February 21, 1958. A tool pusher is the manager in charge of a land-based drilling rig.

44. Ibid.; John Pearson, interview.

45. *PW*, January 24, 1958.

46. *PW*, February 21 and April 4, 1958.

47. Edith Penrose, *The Large International Firm in Developing Countries: The International Petroleum Industry* (Cambridge, MA: MIT Press, 1968), 87, 178–79; Anthony Sampson, *The Seven Sisters: The Great Oil Companies and the World They Shaped* (New York: Bantam Books, 1975), 174. Also see J. E. Hartshorn, *Oil Companies and Governments: An Account of the International Oil Industry and Its Political Environment* (London: Faber & Faber, 1962) and *Oil Trade: Politics and Prospects* (Cambridge: Cambridge University Press, 1993).

48. Terzian, *OPEC*, 4–6, 14–24; Yergin, *The Prize*, 514–15; Hartshorn, *Oil Companies*, 134–35. World demand was expanding by 5 percent annually during the decade, but production rose to 20 million barrels a day.

49. Yergin, *The Prize*, 509; PW, July 18, 1958.

50. *PW*, August 30, 1957, and August 1 and September 26, 1958.

51. Fuad Rouhani to WJ, September 29, 1958, WJP; Hussein Ala to WJ, October 6, 1958, WJP; Manucher Farmanfarmaian to WJ, September 15, 1958, WJP.

52. *PW*, August 8, 1958.

53. *PW*, December 5, 1958; PW, September 19 and November 7, 1958; Macris to Marashian, October 28, 1958, OMP.

54. *PW*, April 19 and 26, 1957; Robert Sherwood, memo to Department of State,

August 29, 1957, 880.2553/8–2957, RM/R, RG 59, NA.

55. *PW*, December 19, 1958; WJ to Marashian, February 5, 1959, OMP.

56. Ian Seymour, *OPEC: Instrument of Change* (London: Macmillan Press, 1980), 20–23.

57. Terzian, *OPEC*, 23–24; Yergin, *The Prize*, 515.

58. Yergin, *The Prize*, 511; Edwin Lieuwen, "The Politics of Energy in Venezuela," in *Latin American Oil Companies and the Politics of Energy* (Lincoln: University of Nebraska Press, 1985), ed. John Wirth, 205; Juan Pablo Perez Alfonso to WJ, September 18, 1956, WJP.

59. Yergin, *The Prize*, 512; Seymour, *OPEC: Instrument*, 19–22.

60. Yergin, *The Prize*, 512–13, 516; Adam Smith, *Paper Money* (New York: Summit Books, 1981), 148–49; *PW*, April 3 and 24, 1959; Terzian, *OPEC*, 23–24.

61. *PW*, June 20, 1958; Nolte, interview; WJ, interview by Yergin; WJ, interview notes re Tariki meeting in 1958 appointment diary, WJP.

62. Perez Alfonso to WJ, May 31 and September 18, 1956, WJP.

63. WJ, interview by Yergin.

64. Jack Hayes, telephone interview by author, June 16, 1995; Brandon Grove, Diary, April 13, 1959, BGP.

65. Grove, diary, April 14 and 16, 1959, BGP; WJ, interview by Yergin; Terzian, *OPEC*, 28. Responding to a quote in Terzian's account, Wanda told Yergin that Tariki remembered incorrectly when he wrote in 1983 that he had wanted to meet Perez Alfonso since 1951. When she told Tariki about the Venezuelan, Tariki was not even clear about his name.

66. Adam Smith, *Paper Money*, 144–46; *Forbes*, July 6, 1981.

6. OPEC'S MIDWIFE

1. Michael Hubbard, memo to BP, April 29, 1959, Deighton File, Middle East Center, Oxford, UK; Richard Nolte, interview by author, February 8, 1996; Georgia Macris, interview by author, February 10, 1995; Feisal Mazidi, interview by author, October 31, 1995; Alirio Parra, interview by author, November 2, 1995.

2. Hubbard, memo to BP; *PW*, May 8, 1959; Onnic Marashian, interview by author, June 4, 1996.

3. A. T. Chisholm, memo to chairman, BP, April 30, 1959, Deighton File; "Record of a Meeting with the British Oil Companies Held in Sir Roger Stevens' Room," Foreign Office, May 13, 1959, FO 371 140378, PRO.

4. *PW*, May 8, 1959; Hubbard, memo to BP.

5. Miles Copeland, *The Game of Nations: The Amorality of Power Politics* (New York: Simon & Schuster, 1969), 243, 257; Daniel Yergin, *The Prize: The Epic Quest for Oil, Money, and Power* (New York: Simon & Schuster, 1991), 508–10; Diane Kunz, *The Economic Diplomacy of the Suez Crisis* (Chapel Hill: University of North Carolina Press, 1991), 45, 59, 184–85; Nathan Citino, *From Arab Nationalism to OPEC: Eisenhower, King Saud, and the Making of U.S.-Saudi Relations* (Bloomington: Indiana University Press, 2002), 150–56. Syria joined with Egypt in 1958 to form the United Arab Republic.

6. John Lewis Gaddis, *Strategies of Containment: A Critical Appraisal of Postwar American National Security Policy* (New York: Oxford University Press, 1982), 181–82; Yergin, *The Prize*, 480–93.

7. Copeland, *Game of Nations*, 182, 225–44; Yergin, *The Prize*, 508–10; Memorandum of Conference with President Eisenhower, July 23, 1958, 98–99, and Memorandum of Discussion, National Security Council, July 24, 1958, U.S. State Department, *Foreign Relations of the United States: Diplomatic Papers* (Washington, DC: Government Printing Office, 1948–90), Vol. 12, 102, 105–8. Said Dulles: "We must regard Arab nationalism as a flood which is running strongly."

8. Copeland, *Game of Nations*, 257; *Fortune*, October 1958.

9. Copeland, *Game of Nations*, 20, 70, 187, 200, 257–59; P. J. Vatikiotis, *Nasser and His Generation* (New York: St. Martin's Press, 1978), 212–14, 231–39, 356–59; Anthony Nutting, *Nasser* (New York: E.P. Dutton, 1972), 254–55.

10. Manucher Farmanfarmaian and Roxane Farmanfarmaian, *Blood and Oil: Memoirs of a Persian Prince* (New York: Random House, 1997), 336, 339; Manucher Farmanfarmaian, interview by author, December 4, 1996; Grove, diary, April 16, 1959, BGP; *PW*, May 8, 1959.

11. *PW*, January 23, February 20, and April 24, 1959; David Hirst, *Oil and Public Opinion in the Middle East* (New York: Praeger, 1966), 76–83; WJ to Marashian, October 21, 1959, OMP; Marashian to WJ, October 29, 1959, OMP; Marashian, telephone interview by author, June 24, 2001.

12. *Newsweek*, April 27, 1959; *Business Week*, May 2, 1959; *Time*, April 27, 1959; *New York Times*, February 15 and April 12, 18, 19, 20, 24, 1959; *Wall Street Journal*, April 16, 17, 20, 22, 24, 1959.

13. *PW*, April 24 and May 8, 1959.

14. Hubbard, memo to BP; *PW*, April 24, 1959.

15. *PW*, May 8, 1959.

16. Ibid.

17. Ibid.; "Work Sessions of the Congress, 18–21 April 1959," OMP; Marashian, e-mail to author, June 23, 2001.

18. Grove, diary, April 20, 1959, BGP; Michael Ameen, interview by author, November 25, 1997; Macris, interview.

19. Ameen, interview; Mazidi, interview; Tammam Yamani, interview by author, November 10, 1996. More than a dozen men said that their wives had fond memories of Wanda, including Rostam Bayandor, Eric Drake, James Hunt, and Denis Wright.

20. Macris, interview, February 10, 1995, and telephone interview by author, October 14, 1996; Youssef Ibrahim, interview by author, November 11, 1996; WJ to Bayandor, March 9, 1961, WJP.

21. James Hunt, interview by author, June 5, 1996; Macris, interview, February 10, 1995; Helen Avati, interview by author, November 2, 1995; Grove, diary, 1959–67, BGP.

22. Farmanfarmaian and Farmanfarmaian, *Blood and Oil*, 339; Farmanfarmaian, interview; John Buckley, interview by author, July 7, 1996.

23. Farmanfarmaian and Farmanfarmaian, *Blood and Oil*, 340–42; Pierre Terzian, *OPEC: The Inside Story*, trans. Michael Pallis (London: Zed Books, 1985), 27–28; Yergin, *The Prize*, 518.

24. Farmanfarmaian and Farmanfarmaian, *Blood and Oil*, 342; Farmanfarmaian, interview.

25. *PW*, May 1, 1959; Terzian, *OPEC*, 28. Although Terzian quotes Tariki as saying that Wanda introduced him to Perez Alfonso in the Shepheards Hotel, Wanda told a number of people, including Yergin, that it was in the Hilton, and Marashian told the author that it was in the Hilton. Wanda also told Yergin that Perez Alfonso would have never said, "So that's you, the one making all the noise," as Tariki reported. Perez Alfonso was a much more formal person, she said. WJ, interview by Yergin, September 11, 1987, DYP.

26. *PW*, May 8, 1959; Anthony Sampson, *The Seven Sisters: The Great Oil Companies and the World They Shaped* (New York: Bantam Books, 1975), 190; Marashian, interview, June 4, 1996.

27. "A Strange New Plan for World Oil," *Fortune*, August 1959: "In the words of one sharp-eyed Arab visitor to these shores, quoted in *PW*, 'Why should the U.S. protect five-barrel-a-day wells that are obsolete? That's like our saying let's not build railroads because what will happen to the poor camel.'" Adam Smith, *Paper Money*, 155.

28. *PW*, May 1, 1959; Marashian, interview, June 30, 2001; WJ appointment diaries, 1957–60, WJP; Grove, diary, 1959, BGP; Robert Mabro, interview by author, October 27, 1995.

29. *PW*, May 1, 1959.

30. *PW*, April 24 and May 1, 1959; Yamani, interview; William Lindenmuth, interview by author, December 7, 1996.

31. K. L. Stock, memorandum, March 25, 1959, POWE 33/2470, PRO; John Pearson, interview by author, June 6, 1996.

32. Sampson, *Seven Sisters*, 167.

33. J. E. Hartshorn, *Oil Companies and Governments: An Account of the International Oil Industry and Its Political Environment* (London: Faber & Faber, 1962), 143, 150–51, 154–55, 178, 388; Jan Nasmyth, interview by author, October 26, 1995; Silvan Robinson, interview by author, November 2, 1995; Dillard Spriggs, interview by author, December 5, 1996; Jack Sunderland, interview by author, August 8, 1995; Penrose, interview; John Bishop, interview by author, May 21, 1997; Buckley, interview; Howard Clark, telephone interview by author, July 19, 1995; Howard Macdonald, interview by author, November 12, 1996.

34. Yergin, *The Prize*, 472–75; Stock, memo, May 13, 1959, FO 371 140378, PRO; Herb Schmertz, interview by author, June 7, 1996.

35. Brandon Grove Jr., interview by author, July 23, 1997; Buckley, interview; James Critchfield, interview by author, August 5, 1995; Spriggs, interview; Sampson, *Seven Sisters*, 14–15; Omar Kamil Haliq to WJ, June 13, 1961, WJP.

36. James Critchfield, interview; C. B. Squire, telephone interview by author, August 29, 1997; Marashian, interview, June 4, 1996; Buckley, interview; Sunderland, interview; Robert D. Kaplan, *The Arabists: The Romance of an American Elite* (New York: Free Press, 1995), 3, 181; Patrick Heren, interviews by author, October 25 and 29, 1995.

37. Brandon Grove Jr., *Behind Embassy Walls: The Life and Times of an American Diplomat* (Columbia: University of Missouri Press, 2005), 28; Grove Jr., interview; Kaplan, *Arabists*, 139, 182–85.

38. Grove Jr., interview; Sunderland, interview; James Critchfield, interview; Napier Collyns, interview by author, August 11, 1995; Ken Miller, interview, February 10, 1995.

39. James Critchfield, interview; WJ appointment diaries, 1956–61, WJP; J. E. Hartshorn, interview by author, November 15, 1995; Lord Christopher Tugendhat, interview by author, November 14, 1996; Mabro, interview; Macdonald, interview; Walter Newton, interview by author, November 25, 1995.

40. Nasmyth, interview; Fred Thackeray, interview by author, November 18, 1996; Hartshorn, interview, October 24, 1995; Tugendhat, interview; Sir David Barran, interview by author, November 15, 1996; Sampson, *Seven Sisters*, 254. The technical *Petroleum Times* and the *Petroleum Press Service* (later the *Petroleum Economist*) did not publish breaking news.

41. Thackeray, interview; Bishop, interview; Ibrahim, interview; John Loudon, telephone interview by author, March 9, 1995; *PIW*, special supplement, April 6, 1992; Sir Eric Drake, interview, November 2, 1995; Farmanfarmaian, *Blood and Oil*, 259. Drake had, according to Farmanfarmaian, unceremoniously fled Iran's Abadan refinery in 1952. Bishop worked for Shell before reporting for *PIW*.

42. Tugendhat, interview.

43. Ibid.

44. Ibid.; Tugendhat, letter to author, November 27, 1996; Hartshorn, interview, October 29, 1995; Robinson, interview; Alfred Munk, interview by author, June 1, 1996; Ibrahim, interview.

45. Munk, interview; Yamani, interview; Andrew Ensor, interview by author, July 15, 1996; Loudon, interview; Joseph Johnston, telephone interview by author, November 7, 1996; Marshall Thomas, interview by author, June 15, 1995; Ken Miller, interview; Henry Moses, telephone interview by author, June 20, 1996. The industry stopped talking with Grove, Moses said.

46. Johnston, interview; Moses, interview; Barran, interview; Lindenmuth, interview. Cash "considered Wanda dangerous," Barran recalled.

47. *PW*, November 6, 1959; C. B. Squire, letter to author, August 7, 1997.

48. Butler, interview; Robert Vitalis, *America's Kingdom: Mythmaking on the Saudi Oil Frontier* (Stanford, CA: Stanford University Press, 2007), 31, 69–70, 117, 121–25, 162, 208–10. Butler had "carte blanche to spend whatever necessary to improve Aramco's image in Saudi Arabia" because Aramco was worried about the potential for nationalization, as in Iran.

49. Michael Weir, British embassy, Washington, memo to Foreign Office, June 17, 1959, FO 371 140378, PRO; Butler, interview; Macris, interview, February 12, 1997.

50. Squire, letter. Squire began to work for McGraw-Hill's *Platt's Oilgram News* in New York in 1958.

51. Squire, letter; Mazidi, interview; Nolte, interview; Jeffrey Robinson, *Yamani: The Inside Story* (London: Simon & Schuster, 1988), 63; William Smith, interview by author, August 8, 1995.

52. Yamani, interview.

53. Mazidi, interview; Alirio Parra, interview by author, November 1, 1995; Munk, interview; Walters, interview; Beverly Lutz, telephone interview by author, May 16, 2006; Barran, interview.

54. *PW*, June 5 and 26, and July 31, 1959. Her reporting picked up the real significance of the warning from Conoco's president, Leonard McCollum—that was not noted in the *New York Times* coverage of June 2, 1959.

55. *PW*, October 16 and 23, 1959; Alirio Parra, interviews by author, November 1 and 2, 1995; Alirio Parra, interview by Yergin, November 11, 1989, DYP; Seymour, *OPEC: Instrument*, 29–32. Perez Alfonso could be stubborn, "as pigheaded as Eric Drake," recalled Parra.

56. Alirio Parra, interviews; *PW*, May 6, 1960. Tariki acknowledged, through Wanda's prodding, that his endorsement of Perez Alfonso's plan was on his own authority, not his government's. Seymour, *OPEC: Instrument*, 31–32.

57. *PW*, February 19, May 27, and July 15, 1960; Howard Page, testimony, March 28, 1974, U.S. Senate, *Multinational Corporations*, part 7, 288; Edith Penrose, *The Large International Firm in Developing Countries: The International Petroleum Industry* (Cambridge, MA: MIT Press, 1968), 195. Also see M. A. Adelman, *The World Petroleum Market* (Baltimore: Johns Hopkins University Press, 1972); Hartshorn, *Oil Companies and Governments*; Bennett Wall, *Growth in a Changing Environment: A History of Standard Oil Company (New Jersey), Exxon Corporation, 1950–1975* (New York: McGraw-Hill, 1988); Yergin, *The Prize*.

58. Wall, *Growth*, 601–2.

59. *Fortune*, January 1963; WJ, interview by Yergin; Yergin, *The Prize*, 520–21; A. Franklyn Williams to Keith L. Stock, June 23, 1958, WJP; Francisco Parra, *Oil Politics: A Modern History of Petroleum* (London: I. B. Taurus, 2005), 97.

60. Denis Wright, memo to Foreign Office, September 1, 1958, EQ1531/15, FO371/133119, PRO; Macris, interview, October 13, 1996; Margaret Clarke, interview by author, October 8, 1996; Denis Wright, interview by author, October 31, 1995. Wanda liked Page's ballroom dancing.

61. Wall, *Growth*, 602.

62. WJ, interview by Yergin; Barran, interview. Although Wanda gave Yergin only a cryptic account of her meeting with Rathbone, she told Sampson that Rathbone was the anonymous executive she quoted. Sampson, *Seven Sisters*, 187n.

63. WJ, interview by Yergin.

64. *PW*, July 22, 1960.

65. *PW*, July 29, 1960.

66. Ibid.

67. John Pearson, interview.

68. Tariki to WJ, August 3, 1960, WJP.

69. Marashian, interview, June 4, 1996; WJ, interview by Yergin.

70. Yergin, *The Prize*, 521–22; Wall, *Growth*, 603; *PW*, August 12, 1960; Ameen, interview. Ameen said that despite this move, "Page, like many top people in the majors, never took the nationalist concerns seriously enough."

71. *PW*, August 19, 1960; Terzian, *OPEC*, 33–34.

72. Ibid.; Farmanfarmaian and Farmanfarmaian, *Blood and Oil*, 343–44.

73. WJ, "Discussion with Entezam," undated, 1960, WJP.

74. Kaplan, *The Arabists*, 200–202; Farmanfarmaian and Farmanfarmaian, *Blood and Oil*, 354–56; Fathollah Naficy, telephone interview by author, June 1, 1996.

75. WJ, "Discussion with Entezam"; Farmanfarmaian, interview; Farmanfarmaian and Farmanfarmaian, *Blood and Oil*, 343. Farmanfarmaian said Wanda asked Entezam specifically about the Gentlemen's Agreement.

76. Ibid.; "Record of Conversation between the Economic Counselor and Miss Wanda Jablonski of 'Petroleum Week' in Tehran on November 28, 1960," British embassy, FO 371/149809, PRO.

77. Farmanfarmaian, interview; Farmanfarmaian and Farmanfarmaian, *Blood and Oil*, 343; WJ to Menzing, November 15, 1960, WJP. Farmanfarmaian, in the only published account of this incident, said that he had heard that Entezam bought the entire print run of the *PW* issue to kill the story, but there is no evidence the issue was canceled or any payment made.

78. Farmanfarmaian and Farmanfarmaian, *Blood and Oil*, 345; Seymour, *OPEC: Instrument*, 36–37; Yergin, *The Prize*, 522–23; Marashian, interview, June 4, 1996; Ameen, interview; *PIW*, February 3, 1992.

7. SMOKING OUT THE SHAH

1. Rostam "Mike" Bayandor, interview by author, November 25, 1995. Bayandor and his wife, Zia, who became close friends of Wanda's, fled to London in 1979. In 1960, Bayandor was technically on the NIOC's public relations staff. National Iranian Oil Company Newsletter, December 1, 1960, 5; *Tehran Journal*, November 22, 1960.

2. Fuad Rouhani to WJ, June 1 and July 21, 1960, WJP; WJ to LeRoy Menzing, November 15, 1960, WJP.

3. *PW*, September 23, 1960; Onnic Marashian, interview by author, June 23, 2001; Ian Skeet, *OPEC: Twenty-Five Years of Prices and Politics* (Cambridge: Cambridge University Press, 1988), 22. Skeet notes that *PW* had "no comment" on OPEC's creation in its September 16 issue, whereas the new *Middle East Economic Survey* did. *PW*, however, was not allowed to cover the conference, and the September 15 announcement was too late for *PW*'s deadline.

4. WJ appointment diary, 1960, WJP; Brandon Grove, diary, 1960, BGP; Arthur

Mills, telephone interview by author, November 16, 1996 (Mills cited his Mobil Oil chauffeur logs); WJ, interview by Yergin, September 11, 1987, DYP.

5. WJ, "Why Parent Companies Should Disassociate Themselves from the Arab Oil Congress," undated manuscript (most likely, early 1961), WJP.

6. *PW*, October 21, 1960.

7. Ibid.; *PW*, August 12 and 19, September 2, 16, 23, and 30, and November 4, 1960; Eugene Norris, telephone interview by author, November 6, 1995; Marashian, interview, June 4, 1996. Juan Pablo Perez Alfonso and Abdullah Tariki did not invite her to Baghdad. Perez Alfonso agreed to talk with Marashian, but only as a reporter for the daily *Oilgram News*, not for *PW*.

8. Nathan J. Citano, *From Arab Nationalism to OPEC: Eisenhower, King Saud, and the Making of U.S.–Saudi Relations* (Bloomington: Indiana University Press, 2002), 156; *PW*, October 21, 1960.

9. Feisal Mazidi, interview by author, October 31, 1995; John Buckley, interview by author, July 7, 1996. When Kuwaiti officials learned that the Saudis had gotten two seats on Aramco's board, they insisted on equal treatment. Within two years of earning his degree, Mazidi became a director on Kuwait Oil's board. "Fraser was so rude," Mazidi recalled.

10. WJ appointment diary, 1960, WJP; Mazidi, interview; *PW*, September 2 and October 28, 1960.

11. Marashian, interview, June 4, 1996; Grove, diary, 1960, BGP.

12. J. E. Hartshorn, interview by author, October 24, 1995; Marashian, interview, June 4, 1996; WJ appointment diary, 1960, WJP; WJ to Menzing. Philby's main assignment was for the *Observer* (London). Wanda, who saw him a number of times in Beirut, was deeply shocked when she learned of his defection. Eleanor Schwartz, interview by author, February 10, 1996.

13. Hartshorn, interview; Francisco Parra, interview by author, July 6, 1996; Denise Regan, interview by author, June 4, 1995; Grove, diary, 1960, BGP; Marashian, interview.

14. WJ to Menzing; Fuad Rouhani to WJ, June 1 and July 21, 1960, WJP.

15. WJ to Menzing.

16. WJ, confidential memo, November 1960, FO 371 149809, PRO; Rostam Bayandor, "The Light Side of Wanda's Adventures," *PIW*, special supplement, April 6, 1992; Ian Seymour, interview by author, November 17, 1996; Bayandor, interview; Farmanfarmaian, interview; Eric Schliemann, telephone interview by author, November 22, 1996; transcript of interview by Pertamina official with Wanda in Jakarta, unattributed translation, August 27, 1974, WJP; Marshall Thomas, interview by author, June 30, 1995. Marshall

Thomas said that, years later, Wanda hanged a large poster of the shah in her
New York office and drew, with a black marker, a bull's-eye on his face.

17. WJ, "Notes on 1½ Hour Interview with Shah of Iran by Wanda Jablonski on
November 23, 1960," undated, WJP.

18. Ibid. At that time, Iran's output was averaging a little more than 1 million
barrels a day, whereas the average for Kuwait was 1.6 million, Saudi Arabia
1.1 million, and Iraq a little less than 1 million.

19. WJ, "Notes"; Sir George Harrison, British embassy, Tehran, to Sir Roger
Stevens, December 1, 1960, FO 371 149809, PRO; *PW*, December 9, 1960.

20. WJ, "Notes."

21. Ibid.

22. Bayandor, interview; Sir Denis Wright, interview by author, October 31, 1995;
economic counselor identified only as "Phillips," British embassy, Tehran,
report to Foreign Office, November 28, 1960, FO 371 149809, PRO.

23. WJ, "Notes."

24. *PW*, December 9, 1960; UPI report, *(New Orleans) Times-Picayune*,
December 9, 1960; "Aspects of Relations between the Oil Consortium
and Iran in 1960," Foreign Office, January 10, 1961, FO 371 157649, PRO;
Economist, January 14, 1961; Skeet, *OPEC: Twenty-Five Years*, 26; Pierre
Terzian, *OPEC: The Inside Story*, trans. Michael Pallis (London: Zed Books,
1985), 52.

25. *PW*, December 16, 1960.

26. Phillips, report to Foreign Office.

27. Geoffrey Harrison to Sir Roger Stevens, Foreign Office, December 1, 1960,
FO 317 149809, PRO. Also see Hartshorn to WJ, December 14, 1960, WJP.

28. *La Esfera* (Caracas), January 10, 1961.

29. *PW*, December 23 and 30, 1960, and January 6, 13, and 20, 1961; *Economist*,
January 14, 1961; Hartshorn to WJ, December 14, 1960, WJP; Ashraf Lutfi to
WJ, January 18, 1961, WJP; H. T. Kemp to WJ, January 19, 1961, WJP.

30. Hartshorn, interview.

31. *PW*, December 23, 1960, and January 6, 1961.

32. *PW*, January 13, 1961; WJ appointment diaries, 1960–61, WJP.

33. J. E. Hartshorn, *Oil Companies and Governments: An Account of the International Oil Industry and Its Political Environment* (London: Faber & Faber,
1962), 309–10; Hartshorn, interview. He dedicated the book to Wanda.

34. Ibid.

35. Hartshorn, interview.

36. *PW*, December 30, 1960.

37. Ibid.

38. Alirio Parra, interview, November 1, 1995, recalled Sayid Omar's poetry; James Akins, interview by author, June 14, 1996; WJ appointment diary, 1960, WJP; Mazidi, interview; *Fortune*, March 1957.

39. WJ appointment diary, 1960, WJP; Hunt, interview; Clarke, interview.

40. *PW*, January 20, 1961.

41. Ibid.

42. *Economist*, January 14, 1961; Terzian, *OPEC*, 51–52. There is no evidence that Wanda was a party to any company effort—directly or indirectly.

43. Saba Habachy to WJ, January 20, 1961, WJP.

44. Harrison to Stevens; Mazidi, interview; Marashian, interview, June 4, 1996; Alirio Parra, interview, November 2, 1995.

45. *PW*, January 27, 1961, and December 30, 1960; Terzian, *OPEC*, 53–54.

46. *PW*, January 27, 1961; Terzian, *OPEC*, 93; Skeet, *OPEC: Twenty-Five Years*, 26.

47. *PW*, January 6, 13, and 27, 1961; Frederick Stephens, cable to Shell Libya re locating WJ, December 28, 1960, WJP; John Pearson, cables to *PW*, January 1961, WJP; WJ appointment diary, 1961, WJP. WJ traveled from Lebanon to Egypt, Libya, Tunis, Algeria, and Morocco before coming down with hepatitis and returning through London to New York.

48. WJ appointment diary 1961, WJP; Grove, diary, 1961, BGP; Francisco Parra, interview. A paucity of evidence and Hartshorn's unwillingness to discuss this aspect of their relationship leaves the subject open-ended.

49. Marashian, "Overhaul," unpublished chapter of history of *Platt's*, 1999, OMP; William Bland, telephone interview by author, March 2, 1998; Harriet Costikyan, interview by author, June 4, 1996; WJ to Earl Turner, March 30, 1961, WJP.

50. Georgia Macris, interview by author, July 20, 1994; Marashian, interview; Costikyan, interview.

51. Ervin DeGraff to Harry Waddell, January 21, 1960, WJP; WJ to Standard Oil Co. (New Jersey), February 2, 1960, WJP; Frederic Ahern, United Corp., to WJ, August 21, 1961, WJP; Costikyan, interview.

52. WJ to Standard Oil Co. ("attention Howard Page"), June 17, 1960, WJP; Ahern to WJ, WJP; Macris, interview, July 20, 1994; Squire, interview, September 8, 2002.

53. WJ, "Notes on 5-Hour Conversation with Mohammad Salman, Head of Arab League's Petroleum Department, on Aug. 16, 1960, in New York City," undated, WJP.

54. WJ, "Why Parent Companies Should Disassociate Themselves from the Arab Oil Congress," undated, WJP.

55. Marashian, "Overhaul"; Nelson Bond to PW employees, April 24, 1961, WJP; Bland, interview; Marashian, interview; Macris, interview.

56. *Garden Journal,* May–June 1963; Mary Jablonski, "Fifty Years of Our Life as I Saw It," undated (about 1960), unpublished essay, WJP; Sister Mary Louise Sullivan, Cabrini College, telephone interview by author, October 24, 1996; Macris, interview; Clarke, interview.

57. Salman to WJ, May 3, 1961; Omar Kamil Haliq to WJ, June 13, 1961; Farrokh Najmabadi to WJ, May 8, 1961; Amir Hoveyda to WJ, May 13, 1961, WJP.

58. Elliott Bell to WJ, April 28, 1961, WJP; WJ to Bell, April 28, 1961, WJP; Hartshorn to WJ, April 27 and July 27, 1961, WJP; Marashian, interview, June 4, 1996; Bland, interview; Macris, telephone interview by author, March 3, 1998; Squire, interview; Squire, letter; Marashian to WJ, September 12, 1961, OMP.

59. Hartshorn to WJ, July 27, 1961, WJP.

60. Clarke, interview; Napier Collyns, interview by author, August 11, 1995.

61. Clarke, interview; Barran, interview.

62. Buckley, interview; Costikyan, interview; Macris, interview, February 10, 1995; Spriggs, interview; Barran, interview; Collyns, interview; anonymous chief executive, interview. Costikyan, a McGraw-Hill circulation expert, helped Wanda set up a circulation system and find a printer. Wanda later thanked her with a red Thunderbird.

63. WJ to Marashian, September 4, 1961, OMP; Macris to Marashian, September 28, 1961, OMP.

64. Buckley, interview; Macris, interview, March 3, 1998; Clarke, interview; *Wall Street Journal,* June 22, 1988.

8. PUBLISHING THE "BIBLE"

1. *PIW,* May 7, 1962; *Middle East Economic Survey,* May 11, 1962; Jean-Jacques Berreby, *Pétroles Informations* (Paris), May 18, 1962; Ade Ponikvar to WJ, May 11 and 25, 1962, WJP; John Buckley, interview by author, July 7, 1996; Alirio Parra, interview by author, November 1, 1995.

2. Buckley, interview; Harriet Costikyan, interview by author, June 4, 1996; *Forbes,* June 1, 1971; Louis Moscatello, "They Foresee the Future and It Works for Them," *Family Business,* May 1990; Harry Alter memo to William Mulli-

gan, Dhahran, January 13, 1962, WEM. Aramco analyzed each issue for accuracy. *Platt's Oilgram News* charged $225 a year for more than 250 issues.

3. Buckley, interview; Costikyan, interview; John Buckley, "Folding, Stuffing, and Stamping the First PIW," *PIW*, special supplement, April 6, 1992.

4. J. E. Hartshorn, interview by author, October 24, 1995; Buckley, interview. See, for instance, *PIW*, November 9, 1964.

5. Buckley, interview; *PIW*, April 16, 1962.

6. Jeffrey Robinson, *Yamani: The Inside Story* (London: Simon & Schuster, 1988), 195–96; Margaret Clarke, interview by author, October 8, 1996; Sheikh Ahmed Zaki Yamani, interview by author, November 10, 1996; Ian Seymour, interview by author, November 17, 1996.

7. *PIW*, April 16 and 23, 1962.

8. Sir David Barran, letter to author, May 28, 1996.

9. WJ to New York staff, April 14, 1962, WJP; WJ appointment diary, 1962, WJP.

10. WJ to New York staff, May 7, 18, and June 8, 1962, WJP.

11. WJ to New York staff, April 24, May 7, 22, and June 8, 13, 1962, WJP.

12. WJ to New York staff, May 7, 10, (undated), and June 2, 4, 5, 8, 1962, WJP.

13. WJ to New York staff, June 7, 1962, WJP; Helen Avati, interview by author, October 31, 1995.

14. Georgia Macris, interview by author, October 13, 1996; Clarke, interview; Buckley, interview; Marshall Thomas, interview by author, June 6, 1995; WJ to staff, June 8, 1962, WJP.

15. *PIW*, May 7, 1962; Berreby, *Pétroles Informations*, May 18, 1962; Ponikvar to WJ; Buckley, interview; Francisco Parra, interview by author, July 6, 1996. According to Berreby, Rouhani said an oil company leaked the report to Wanda.

16. *PIW*, February 2, April 16 and 23, August 13, and December 17, 1962; Alirio Parra, interview.

17. *PIW*, April 15, 1963; WJ appointment diary, 1963, WJP; William Randol, interview by author, February 9, 1996.

18. WJ to Ade Ponikvar, June 13, 1962; WJ appointment diary, 1962, WJP.

19. *PIW*, March 19 and July 9, 1962; Pierre Terzian, *OPEC: The Inside Story*, trans. Michael Pallis (London: Zed Books, 1985), 90–91; Robert Vitalis, *America's Kingdom: Mythmaking on the Saudi Oil Frontier* (Stanford, CA: Stanford University Press, 2007), 219–23.

20. *PIW*, July 9, 1962; Clarke, interview.

21. *PIW*, October 22, 1962; Buckley, interview; *Financial Times*, October 30, 1962; *Carta Semanal* (Caracas), September 29, 1962; WJ to Buckley, December 22, 1963.

22. Parviz Mina, interview by author, November 15, 1996; James Critchfield, interview by author, August 5, 1995; John Lichtblau, interview by author, February 11, 1995; James Akins, interview by author, June 14, 1996. Critchfield said that the majors immediately met with officials of the CIA and other federal agencies to voice their alarm over OPEC. John Lichtblau, an industry consultant, was prevented from meeting OPEC officials in Geneva in 1965. Akins said it was official U.S. policy in the 1960s to ignore OPEC.

23. "Minutes of Meeting of Oil Company Representatives," semiannual, 1962–65, quarterly, 1965–1973, Oil 2:24–31, JJM; "Memorandum for Mr. McCloy re International Oil Retainer Fee," 1966, Oil 1:34, JJM; R. G. Follis to John J. McCloy, April 29, 1966, Oil 2:31, JJM; James Critchfield, interview; Napier Collyns, interview by author, August 11, 1995; Barran, interview; Andrew Ensor, interview by author, July 15, 1996; Kai Bird, *The Chairman: John J. McCloy, The Making of the American Establishment* (New York: Simon & Schuster, 1992), 517–18, 732n; Daniel Yergin, *The Prize: The Epic Quest for Oil, Money, and Power* (New York: Simon & Schuster, 1991), 474–75, 584–85; Anthony Sampson, *The Seven Sisters: The Great Oil Companies and the World They Shaped* (New York: Bantam Books, 1975), 242–43, 254–55; Bennett Wall, *Growth in a Changing Environment: A History of Standard Oil Company (New Jersey), Exxon Corporation, 1950–1975* (New York: McGraw-Hill, 1988), 342–44, 758–60. Wall does not report how extensive and frequent these meetings were during the 1960s. Details of the secret meetings emerged from the opening of the McCloy Papers, which, as confirmed by Barran and Critchfield, make it clear that OPEC was of far greater concern to the Seven Sisters than was Russian oil.

24. Critchfield, interview; Buckley, interview; Macris, interview; WJ appointment diaries, WJP; Oil 2:24–31, JJM. Buckley and Macris said that they do not recall knowing anything about the regular McCloy group meetings.

25. Anonymous, interview by author.

26. Ibid.; Hartshorn, interview; Francisco Parra, *Oil Politics: A Modern History of Petroleum* (London: I.B. Taurus, 2004), 103.

27. A. T. Lamb, confidential memo, July 15, 1964, POWE 61/297, PRO; W. C. C. Rose, confidential memo, July 17, 1964, POWE 61/297, PRO. Fallah was "the oil companies' agent at the OPEC meeting."

28. Lamb, memo; Geoffrey Harrison, Foreign Office to M. Stevenson, Ministry of Power, August 5, 1964; Ministry of Power, "Confidential Note of Meeting: OPEC," July 3, 1964, POWE 61/297, PRO; Ian Seymour, *OPEC: Instrument of Change* (London: Macmillan, 1980), 49. That Fallah was an agent for

BP remained secret, but his visit became public. U.S. embassy, Tehran, to Department of State, November 21, 1963, 506 PET 3 OPEC, NA, stated that Fallah was "jealous and resentful of Rouhani."

29. Sir Eric Drake, interview by author, November 2, 1995. Drake said that Fallah was "something of a traitor [to Iran]." But he admired Fallah's toughness. When Fallah was young, Drake said, "the Shah's father hit him once when he was standing with his hands behind his back." Although he landed in jail for his discourtesy, Fallah eventually became Drake's education manager at Abadan. Sir Denis Wright, interview by author, October 31, 1995. References to *PIW* scoops: *Economist*, January 5, 1963; *Financial Times*, July 28 and October 19, 1963, and January 19, 1965.

30. Avati, interview; Buckley, interview.

31. Ibid.

32. Ibid.

33. WJ to Buckley, December 22, 1963, WJP; Andrew Ensor, interview, July 15, 1996; Clarke, interview; Buckley, interview; Thomas, interview; Sir Peter Walters, interview by author, November 14, 1996; Anna Rubino, "Recollections of Life at PIW through the Years," *PIW*, special supplement, April 6, 1992. Wanda wrote Buckley, "I consider this Gulf thing a personal insult by Whitehead and Craig."

34. Buckley, interview; WJ to Buckley, December 22, 1963, WJP; Thomas, interview.

35. Walters, interview.

36. William Smith, interview by author, August 8, 1995; James Tanner, interview by author, December 10, 1997; Hartshorn, interview; Lord Christopher Tugendhat, interview by author, November 14, 1996; for instance, the *Financial Times* of October 30, 1962, January 29, February 20, and October 19 and 29, 1963.

37. Terzian, *OPEC*, 90–91; WJ to Ponikvar, June 13, 1962, WJP; John Bishop, telephone interview by author, May 21, 1997. Even in the late 1960s, Tariki would visit Wanda at her apartment where they played chess while talking about oil.

38. *PIW*, May 13, 1963. Aside from the sheer volume of *PIW*'s OPEC coverage, the majors objected to *PIW*'s publishing lengthy interviews or speeches by the nationalists.

39. Yamani, interview.

40. Ibid.; Drake, interview.

41. *PIW*, November 11, 1963, supplement, and November 18, 1963.

42. Mina, interview; Farrokh Najmabadi, interview by author, June 14, 1996; Rostam Bayandor, interview by author, November 25, 1995; *PIW*, November 5 and December 17, 1962, June 24 and July 15, 1963, and March 16 and December 14, 1964.

43. Hoveyda to WJ, February 20, 1965, WJP; *PIW*, May 17, 1965; James W. Swinhart, U.S. embassy, Tehran, to Department of State, May 6, 1965, PET 6 Iran, NA; L. Milner Dunn, U.S. embassy, Tehran, to State, June 1, 1965, PET 6 Iran, NA; U.S. embassy, Tehran, to State, July 15, 1965, PET 6 Iran, NA. Hoveyda was executed during the 1979 revolution. Bayandor, interview. "Hoveyda got on so well with Wanda," Bayandor recalled.

44. Bayandor, "The Light Side of Wanda's Adventures," *PIW*, special supplement, April 6, 1992; *Independent* (London), March 23, 1992; *Keyhan International* (Tehran), December 28, 1966, and October 30, 1967. *Keyhan* complained that *PIW* published NIOC's agreement with Saudi Arabia two months before it was publicly released.

45. *PIW*, March 22 and 29, 1965; Loudon to McCloy, January 15, 1965, JJM.

46. *PIW*, March 28, 1966. KNPC, still in its startup phase, was not mentioned.

47. WJ diaries, 1966, WJP; U.S. embassy, Kuwait, to Department of State, May 4, 1965, PET 3 OPEC/PET 6 Kuwait, NA. Wanda, "an old friend" of Mazidi's, "was unwilling to show us documents because she was sworn to secrecy."

48. Mina, interview; Ambassador Howard Cottam to Department of State, "Memorandum of Conversation," March 3, 1966, NA; cables, March 7, April 13, and July 12, 1966, PET 3 OPEC/PET 15–2 Kuwait, PET 6 Kuwait, NA.

49. Yamani, interview; Mina, interview; Ameen, interview; Lichtblau, interview; Mazidi, interview; Barran, letter to author, May 28, 1996. "Wanda did just what the oil executives did not want her to do. She wrote about OPEC almost every week, whereas the general press ignored [it]," Lichtblau said.

50. Ian Skeet, *OPEC: Twenty-Five Years of Prices and Politics* (Cambridge: Cambridge University Press, 1988), 57; Yergin, *The Prize*, 555–58; Sampson, *Seven Sisters*, 209–11; Thomas, e-mail to author, February 22, 2007.

51. Alirio Parra, interview; Nasmyth, interview; Penrose, interview; Adelman, interview; Barran, interview; Drake, interview; *PIW*, January 18, 1965; Skeet, *OPEC: Twenty-five Years*, 32; Francisco Parra, *Oil Politics*, 107. Nasmyth and Drake said OPEC was not the "sideshow" that Skeet said it was. That OPEC kept the majors from lowering posted prices, despite an oversupply, was significant.

52. Yergin, *The Prize*, 557; Clarke, interview; Ken Miller, interview by author, February 11, 1995; Buckley, interview; PW, January 13 and 20, 1961; *PIW*,

December 9, 1963, and April 19, 1965; WJ correspondence, November 1963–February 1964, WJP; Buckley to WJ, April 14, 1965, WJP; WJ appointment diaries, 1960–65, WJP.

53. Ken Miller, interview; Mina, interview; *PIW*, April 19, 1965; WJ appointment diaries, 1965, WJP; T. A. Hillyard to WJ, March 2, 1965, WJP; WJ-Buckley correspondence February–April 1965, WJP.

54. John Tinny, U.S. embassy, Tripoli, to Department of State, September 30, 1965, F76092–0406, NA; Ken Miller, interview; Ken Miller to Wanda and John Buckley, January 18, 1966, WJP. Wanda's classic way of getting people to talk was to imply that she knew more than she could say.

55. Ken Miller, interview; Macris, interview; Buckley, interview; Leyna Miller, telephone interview by author, July 16, 2006.

56. *PIW*, June 7 and 21, 1965, August 4, 1969, and May 18, 1970. Libyan output by 1969 exceeded 2.9 million barrels a day.

57. *PIW*, February 10 and April 7, 1969; David Newsom, U.S. embassy, Tripoli, to Department of State, February 24, 1969, PET 2 Libya, and April 8, 1969, PET 10–3 Libya, NA; Yergin, *The Prize*, 576–79. Leyna Miller was beaten, and her glasses broken during the coup. Ken Miller was in New York and was allowed back in the country two weeks later. He was jailed for a day, then released, and he and his wife left for Rome two days later. Within two weeks, they were allowed back in Libya, where Ken Miller continued to report for *PIW* for 18 months. Leyna Miller, interview.

58. *PIW*, September 14 and October 12, 1970; Seymour, *OPEC: Instrument*, 69–70; Yergin, *The Prize*, 576–79.

59. *PIW*, January 25, 1971; Sampson, *Seven Sisters*, 344. Historian Abbas Amanat underscored the current implications of this psychological issue in "The Persian Complex," *New York Times*, May 25, 2006: "For a country like the United States that is built on paradigms of progress and pragmatism, grasping the mythical and psychological dimensions of defeat and deprivation at the hands of foreigners is difficult. Yet the Iranian collective memory is infused with such themes."

60. Ameen, interview; Rawleigh Warner, telephone interview by author, June 11, 1996; Walters, interview; Tanner, interview; *New York Times*, September 24, 1969. Ameen saw copies of *PIW* "on everyone's desk." Warner said, "When I was CEO [of Mobil], I used to read *PIW* every week as soon as it came in. Everyone did."

61. Hartshorn, interview; Buckley, interview; Avati, interview; WJ to Buckley, December 22, 1963, WJP. Hartshorn, who had little interest in financial matters, said he simply returned the legal document to her.

62. Buckley, interview; Macris, interview; John Bishop, telephone interview by author, July 6, 1997; Bishop, letter to author, undated, June 1997; Macris, interview, October 14, 1996. According to Macris, Wanda's lawyer, Ted Kupferman, also had some small shares in *PIW*, which he apparently later returned. Wanda also insisted that Buckley return the small stake that he held in *PIW*, which he did in a swap for title to property she owned in Montauk.

63. Buckley, interview; Bishop, interview; Macris, interview, February 12, 1997. WJ was forever anxious over money. She refused to have more than $10,000 in one savings account, so Buckley had to deposit subscription money in fifteen accounts. "She always worried that she would lose *PIW*," Macris recalled.

64. Thomas, interview; Clarke, interview.

65. Joe Roeber, interview by author, October 30, 1995; Macris, telephone interview by author, February 12, 1997; Clarke, interview; Costikyan, interview; Schwartz, interview. The death of one noteworthy friend, Oyster Bay neighbor and *Vogue* editor Ilka Chase ("It was one-upsmanship all the time when they were together," remembered Costikyan), however, was so painful that Wanda said she would no longer make friends because they were too hard to lose.

66. Ameen, interview; Macris, interview by author, October 14, 1996; Clarke, interview; Costikyan, interview; Critchfield, interview; Avati, interview; Hartshorn, interview; Beverly Lutz (formerly Parra), telephone interview by author, May 17, 2006; Bayandor, interview.

67. WJ appointment diaries, 1967–69, WJP; Brandon Grove Jr., interview by author, July 23, 1997. Brandon Grove, diary, 1966–69.

68. WJ appointment diaries, 1967, 1976, and 1978, WJP; Alirio Parra, interview; Clarke, interview; Leyna Miller, interview; Nasmyth, interview; Eric Schliemann, telephone interview by author, November 22, 1996; Eugene Norris, telephone interview by author, November 6, 1996; Schliemann to WJ, July 14, 1967, WJP.

69. Costikyan, interview; Regan, interview; Yamani, interview; Thomas, interview; Bayandor, interview. All recall Wanda's saying McFadzean had proposed. Michael Saltser, interview by author, June 4, 1995; Lichtblau, interview; Clarke, interview; Macris, telephone interview by author, July 22, 1995; Hunt, interview; Dr. Michael Osborne, interview by author, December 6, 1996; Collyns, interview. When Wanda died in 1992, some obituaries said that her three marriages had ended in divorce, but there are no records of any marriage other than her twelve-year marriage to John Jaqua.

70. Wright, interview; Hunt, interview.

71. Costikyan, interview; Schwartz, interview; Lutz, interview; Macris, interview; Thomas, interview; Hunt, interview; Clarke, interview; Collyns, interview.

72. Macris, interview; Buckley, interview.

73. Yamani, interview; Alirio Parra, interview; Ensor, interview; Barran, interview; Ameen, interview; Buckley, interview; Lois Critchfield, letter to author, March 9, 1998. Also see *PIW*, 1965–72; U.S. embassy, Tripoli, to Secretary of State, March 21, 1966, PET 6, NA; U.S. embassy, London, to Secretary of State, August 10, 1967, PET 2, NA; Sojuznefteexport to Editor, *PIW*, April 19, 1971, WJP. Lois Critchfield, CIA analyst, said: "I personally know that *PIW* was a major resource document . . . and distributed widely" at the CIA. Sampson, *Seven Sisters*, 345. By the mid-1960s, the Russians had seven *PIW* subscriptions.

9. FINAL CONFRONTATIONS

1. Georgia Macris, interview by author, October 13, 1996; Helen Avati, interview by author, October 31, 1995; Margaret Clarke, interview by author, October 8, 1996.

2. Ibid.; William Randol, interview by author, February 9, 1996; Alfred Munk, interview by author, June 1, 1996.

3. Daniel Yergin, *The Prize: The Epic Quest for Oil, Money, and Power* (New York: Simon & Schuster, 1991), 583–84; Ian Seymour, *OPEC: Instrument of Change* (London: Macmillan Press, 1980), 221–25; "Minutes of OPEC/Gulf Group Meeting Held in Riyadh — 7 May 1972," Oil 3:6, JJM.

4. *PIW*, March 20 and 27, April 3 and 24, and July 17, 1972.

5. Anthony Sampson, *The Seven Sisters: The Great Oil Companies and the World They Shaped* (New York: Bantam Books, 1975), 277–83; Sheikh Ahmed Zaki Yamani, interview by author, November 10, 1996; Bennett Wall, *Growth in a Changing Environment: A History of Standard Oil Company (New Jersey), Exxon Corporation, 1950–1975* (New York: McGraw-Hill, 1988), 833; Herb Schmertz, interview by author, June 7, 1996; James Akins to Department of State, October 5, 1972, 182567, PET 3 OPEC, NA. Schmertz said, "We told the U.S. government as little as possible."

6. *PIW*, October 9, 1972; Yergin, *The Prize*, 584–85.

7. Ibid. The Gulf States' participation share was scheduled to rise to 51 percent by 1983.

8. *New York Times*, August 19, 1979; Avati, interview; Macris, interview. The obligation to inform went back to the Red-Line Agreement of 1928, which

carved up the oil reserves of the old Ottoman Empire. Patrick Heren's 1992 obituary of Wanda in *PIW* (February 3, 1992) erroneously attributed the scoop to her, based on what Wanda had told him some years earlier.

9. Randol, interview.

10. *Economist*, October 14, 1972.

11. Macris, interview; Ian Seymour, interview by author, November 17, 1996.

12. Ibid.; WJ, "Trip Expenses," 1972, WJP.

13. Macris, interview; Macris cable to WJ, undated 1972, WJP; Peter Walters, interview by author, November 14, 1996.

14. Seymour, interview; Macris, interview.

15. Yamani, interview.

16. Seymour, interview; Macris, interview; WJ to Wadad and Mahmoud Makhlouf, January 17, 1973, WJP; WJ, "Trip Expenses," 1972, WJP.

17. James Critchfield, interview by author, August 5, 1995; Yamani, interview; Macris, interview; WJ, "Notes," undated, WJP; WJ appointment diary, 1972, WJP; WJ, "Travel Expenses," 1972–73, WJP.

18. Robert Blair Kaiser, *Clerical Error: A True Story* (New York: Continuum, 2002); *Independent* (London), August 6, 1999; *Observer* (London), March 17, 2002; *Los Angeles Times*, July 31, 1999; *New York Times*, July 30, 1999; *Washington Post*, March 27, 1987.

19. *PIW*, November 13 and 20, 1972; WJ to H. Eugene Bird (U.S. embassy, Jeddah), January 31, 1973.

20. "Itinerary and Notes," 1972, WJP; *PIW*, October 23 and 30, November 13 and 20, and December 4 and 25, 1972; WJ cable to Macris, undated, November 1972, WJP; Macris to WJ, November 24, 1972, WJP.

21. *PIW*, November 27 and December 11 and 18, 1972; WJ appointment diaries, 1965–75, WJP; Clarke, interview; John Bishop, telephone interview by author, May 21, 1997; Marshall Thomas, interview by author, June 6, 1995; WJ to Mana Said al-Oteiba, January 30, 1973, WJP.

22. Macris, interview; WJ, "Travel File," 1972–73, WJP; Thomas, interview; James Tanner, interview by author, December 10, 1997. Because of Piercy's ban on *PIW*, some Exxon employees subscribed by using personal checks.

23. Macris, interview; Thomas, interview; Randol, interview; Marshall Thomas, untitled note, *PIW*, special supplement, April 6, 1992.

24. *New York Times*, December 21, 1977; *PIW*, April 23, 1973; James Critchfield, interview. Senate Select Committee on Intelligence, news release on staff study, "U.S. Intelligence Analysis and the Oil Issue, 1973–74," December 20, 1977. Release cites *PIW*, the *Financial Times*, and the *Wall Street Journal* as the three specialized public sources that "reported more consistently on

changing intentions of the Saudis" in 1973 than did the U.S. intelligence services. Critchfield's reports, however, were separate from the daily intelligence briefings.

25. *PIW*, October 22, 1973; Yergin, *The Prize*, 606–8.

26. James Critchfield, interview; Thomas, interview; Clarke, interview; Lois Critchfield, letter to author, July 26, 2006, and e-mail to author, July 1, 2007. James Critchfield was chief of the CIA's Near East Division from 1960 to 1968 and then became the CIA's national intelligence officer for energy. Wanda's cooperation was voluntary.

27. Rawleigh Warner, telephone interview by author, June 11, 1996; Schmertz, interview; Thomas, interview.

28. Yamani, interview.

29. Ibid.; Andrew Ensor, interview by author, June 15, 1996; Warner, interview.

30. Thomas, interview; Mai Yamani, interview by author, November 10, 1996.

31. *Christian Science Monitor*, December 7, 1973; Thomas, interview.

32. *New York Times*, February 5, 1977; Liliano Africano, "When She Talks about Oil, People in High Places Listen," Liliano Africano Princeton Features, 1977, WJ clippings file, WJP; *Frau im Leben*, August 1977.

33. *CBS Evening News*, October 23, 1977, Burrelle's TV Clips, WJP.

34. Ibid.

35. Macris, interview; Clarke, interview; Thomas, interview; Louis Moscatello, telephone interview by author, October 8, 1997; John Lichtblau, interview by author, February 11, 1995; Diane Munro, interview by author, June 9, 2006.

36. Barth Healey, interview by author, October 15, 1996; Moscatello, interview; Tanner, interview; Larry Goldstein, interview by author, October 15, 1996.

37. Ibid.; Thomas, interview; Joe Roeber, interview by author, October 30, 1995; Yergin, *The Prize*, 721–26.

38. Kenneth Miller, interview by author, February 10, 1995; Thomas, interview; Clarke, interview; Moscatello, interview; Macris, interview.

39. David Wallenchinsky, "A Newsletter on Newsletters," 1975, www.trivia-library.com/a/history-of-marketing-and-newsletters; Thomas, interview; Jan Nasmyth, interview by author, October 26, 1995; Thomas Wallin, interview by author, August 9, 1995; "Petroflash File," MTP; WJ, Deposition, August 2, 1988, U.S. District Court, Southern District of New York, 88 Civ. 3840, 155, 163. Wanda eventually supported Thomas's joint venture with New Jersey-based *Oil Buyers' Guide*, but when McGraw-Hill bought *Oil Buyers' Guide*, it also bought out Wanda's share in "Petroflash," which she split with Thomas. She then let him put out a monthly markets supplement to the newsletter, but it was not timely.

40. Macris, interview; Thomas, interview; Miller, interview; WJ, Deposition, 145, 165, 372, 597; Thomas e-mail to author, February 27, 2007.

41. Macris, interview; Thomas, interview; Wallin, interview; WJ, Deposition, 180, 383, 388, 398–401, 456–57, 790. In 1978, Wanda established a trust to ensure the continuation of *PIW* in perpetuity, with several personal friends as trustees, replaced in 1982 by her accountant, her lawyer, and John Treat. In mid-1985, she revoked that agreement.

42. "Business Plan," September 1986, MTP; Thomas, interview; Wallin, interview; Munro, interview; Guy DiRoma, interview by author, June 9, 2006; Thomas and Wallin to prospective investors, April 28, 1986, MTP; Thomas to Adrian Binks, May 5, 1987, MTP; Thomas to John Treat, November 11, 1985, MTP; WJ, Deposition, 8, 456.

43. WJ, Deposition, August 2, 1988, 8–9, 18–19, 29, 88–89, 178, 312, 348, 562–64; DiRoma, interview.

44. WJ, Deposition, 55–62, 68; *New York Times*, June 6, 1988; Kenneth Miller, memo on employee benefits, May 2, 1988, MTP; *Wall Street Journal*, June 22, 1988; UPI, June 2, 1988; Reuters, June 3, 1988; *New York Times*, June 6, 1988; Wallin, interview; DiRoma, interview. Kim Fuad did not quit, citing personal reasons.

45. *Wall Street Journal*, June 22, 1988.

46. Wallin, interview; Michael Saltser, interview by author, June 4, 1995; order, U.S. District Court, Southern District of New York, 88 Civ. 3840, June 8, 1988; *New York Times*, June 7, 1988; Reuters, July 7 and November 11, 1988. A November 10, 1988, court order banning weekly or biweekly facsimile transmissions was vacated January 24, 1989, when the court determined that lifting the restraints would not cause *PIW* "irreparable harm."

47. Edward Morse, "A Tribute to *PIW*'s Founder, Wanda Jablonski," *PIW*, special supplement, April 6, 1992; *Wall Street Journal*, June 22, 1988; Wallin, interview; Thomas, interview.

48. Dr. Michael Osborne, interview by author, December 6, 1996; *New York Times*, May 14, 1991.

49. Vladimir Krcmery, interview by author, July 9, 2003; Osborne, interview; Vladimir Krcmery, letter to author, September 19, 1995; WJ appointment diaries, 1989–90, WJP; Bayandor, "The Light Side of Wanda's Adventures," *PIW*, special supplement, April 6, 1992.

50. Osborne, interview; Wallin, interview.

51. Saltser, interview; Osborne, interview; Denise Regan, interview by author, June 4, 1995. Wanda left a small New York cooperative apartment to Regan.

52. Yergin, *The Prize*, 748–51; *PIW*, July 14, 1986.

53. Ibid.; *Wall Street Journal,* July 15, 1986.

54. Parra, *Oil Politics,* 286–87; *Wall Street Journal,* May 14, 1985; *PIW,* July 23 and 30, 1984, and March 11, 1985.

55. *New York Times,* October 30 and 31, 1986; *Financial Times,* October 30 and 31, 1986; *Wall Street Journal,* November 3, 1986; *Economist,* November 8, 1986; Yergin, *The Prize,* 763.

56. Yamani, interview; Ali Khalifa al-Sabah, e-mail to author, December 13, 2006.

57. Yamani, interview; Robert Mabro, interview by author, October 27, 1995. Mabro said, "Yamani got a call through to Wanda about his firing because *PIW* is an American paper and Yamani wanted the protection of the U.S. Nobody would have noticed it if [the story] had been in [the *Middle East Economic Survey*]. This was very important to Yamani. In the end he needed Wanda much more than *MEES.*"

58. Ibid.; *PIW,* November 24, 1986; Wallin, interview. *PIW* reported that the source "had arrived in New York last week," but Wallin suspects Wanda wrote that to deflect attention from the real source, who was definitely calling from abroad.

59. Ibid.

60. Yamani, interview.

61. Tanner, interview; Khalifa, e-mail.

62. Tanner, interview.

63. Robert Mabro, "Wanda in Her Later Years: Cynicism with a Soul," *PIW,* special supplement, April 6, 1992.

64. WJ to LeRoy Menzing, December 24, 1956, WJP.

65. J. E. Hartshorn, interview by author, October 24, 1995; Lord Christopher Tugendhat, interview by author, November 14, 1996; Walter Newton, interview by author, November 25, 1995; Nasmyth, interview; Walter Levy, interview by author, August 9, 1995; Tanner, interview; "Wanda Jablonski," TJFR/MasterCard International's 100 Business News Luminaries of the Century, December 21, 1999, M2 Presswire.

Index

Page numbers in *italics* refer to photographs.